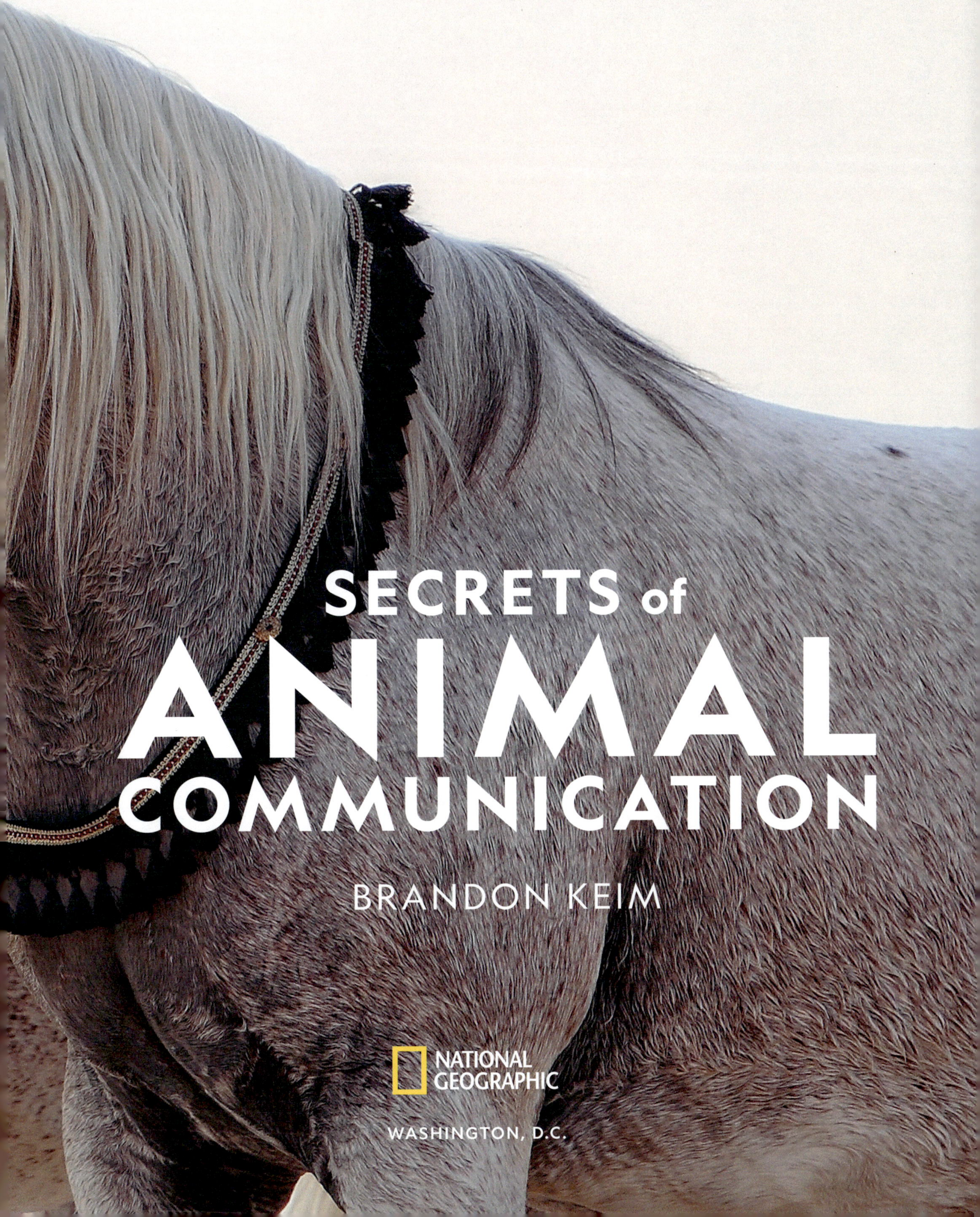

SECRETS of ANIMAL COMMUNICATION

BRANDON KEIM

NATIONAL GEOGRAPHIC

WASHINGTON, D.C.

CONTENTS

THAT HUMAN LANGUAGE is such an exceptional form of communication can sometimes obscure the richness of other ways that messages are shared. From two horse friends nuzzling (previous pages) to the synchronized flashes of fireflies (left), the world is full of meaning.

INTRODUCTION

A BROADER SPECTRUM

Until well into the 20th century, human languages were considered the highest possible form of communication. Language was said to "make us human," and scientific traditions that viewed other animals as simpleminded also regarded their communications as unsophisticated. However lovely a bird's song might sound to our ears, it wasn't thought to be especially meaningful.

Nowadays scientists are much more open-minded. Research journals overflow with insights into animal thoughts and feelings. And just as our own intelligence is newly understood as one point in a fantastic spectrum rather than the pinnacle of an evolutionary hierarchy, a new awareness of animal communications has flourished, too.

Rather than comparing them to our own undeniably extraordinary languages, and then dismissing them as lacking in comparison, we can simply appreciate animal vocalizations—as well as animals' gestures and scents and even, in the case of cuttlefish, patterns rippling across their skin—as myriad, marvelous ways of exchanging information.

That these communications are not human language is just a fact, not a judgment. It's also worth reflecting on how much meaning can be conveyed with a mere look or gesture. For all the significance attached to language, many of our own most important messages don't take linguistic forms.

From that perspective, the world is revealed as a place of constant communication. Dogs leaving their mark on neighborhood smelling posts, ants waving antennae, the chirps of songbirds staying in contact at a distance—signals and meanings are ubiquitous. Even the spots on a butterfly or a strawberry's eye-catching coloration can be understood as parts of dialogues exchanged across evolutionary time.

With such awareness comes a richer sense of the world we share. It can reveal problems—ship noise makes it hard for whales to hear one another; chemical pollution can impair the cognition that underlies communication—and nourish a deeper kinship with other creatures. Communication is the essence of relationships, human and nonhuman alike.

We might not be able to translate the back-and-forth between a mother bear and her cub, yet at a fundamental level we understand them. And if the notion that language makes us human and unique feels dated, perhaps it's time for a different claim. Animal communication is everywhere, but humans are unique in our intellectual quest to understand it.

EMPEROR PENGUINS (opposite) have vocal tracts that let them speak with two voices at once. That makes it easier for them to recognize one another's calls.

PEOPLE DON'T often think of hermit crabs (below) as having much to say. Yet they too communicate and even negotiate mutually beneficial exchanges of shells.

GREAT MINDS

One expanded our imaginations; the other, our science. Together they anticipated a new appreciation of animal communication.

• HUGH LOFTING (1886-1947)

Although the human carnage of World War I would influence a generation of artists, English author Hugh Lofting was also moved by its toll on animals. "However seriously a soldier was wounded, his life was not despaired of," wrote Lofting of his wartime experiences, yet that soldier's wounded horse would be "put out by a timely bullet."

Lofting transmuted that trauma into the character of Doctor Dolittle, a kindhearted Victorian veterinarian who spoke to animals in their own languages and even treated them as equals. Beloved by generations of children and their parents, the stories took the form of innocuous genre fiction yet contained some very powerful ideas.

At a time when animal welfare was still in its social infancy, Lofting's protagonist condemned bullfighting and foxhunting and the capture of wild animals for entertainment. And though mainstream science would soon deny animal intelligence, Doctor Dolittle suggested that animals had much to say—and that, if we kept an open mind, we could even come to understand them.

• RICHARD GARNER (1848-1920)

In 1877, the inventor Thomas Edison devised the first phonograph, a device that recorded sound vibrations as etchings on a foil-wrapped cylinder. By amplifying the vibrations of a needle moving across those etchings, the original sound could be reproduced.

The technology eventually transformed how music is produced, but schoolteacher turned primatologist Richard Garner had other ideas. A devotee of Charles Darwin's then young theory of evolution, Garner hypothesized that primate communication gave rise to human language. Using one of Edison's newfangled devices, Garner recorded the chatter of monkeys and great apes.

He broke their vocalizations down into what he regarded as the primal antecedents of human language; he also played the recordings back to them, and by observing their reactions tried to infer the calls' meanings.

Though his findings fell into disfavor among academics skeptical that primates could speak and think in sophisticated ways, Garner's ingenious experimental methods would be used—and his insights validated—nearly a century later in pioneering studies of monkey communication.

"Their speech is the only gateway to their minds," wrote Garner, "and through it we must pass if we would learn their secret thoughts and measure the distance from mind to mind."

BECAUSE SPOKEN language is so essential to us, people pay more attention to vocalizations than gestures. Yet to dogs (opposite), movements are full of meaning.

PART ONE

WORLDS OF COMMUNICATION

What is communication? It's a seemingly simple question, yet one that scientists and philosophers and scholars have debated at length. There is no universally accepted definition, but instead many ways of thinking about it.

Some researchers have argued that communication requires the individuals involved to possess "theory of mind," or the ability to reason in complex ways about what others think. Humans possess this ability in spades. Few other species seem to, though, and by this definition animal communication would be a rare thing indeed.

Or perhaps communication doesn't demand a theory of mind but does require intentionality. By that light, a monkey who cries out in alarm at the sight of an approaching predator isn't actually communicating unless she intends to warn her companions.

It's also been asserted, though, that intentionality isn't so important. What matters is that a signal leads to a change in behavior. To continue the monkey example, a frightened yelp is communication so long as her companions react.

Communication might even be conceived more broadly: as anything that's evolved to transmit information, with "information" defined as a reduction of uncertainty. Understood in this way, communication is found in the speckles on a trout's flank and the courtship displays of whooping cranes, the abdomen-wagging dances of honeybees and—yes—monkey alarm calls.

That's the understanding reflected in these pages. Cues, signals, messages—communication is everywhere, and life is a constant gathering of information.

THE PATTERNS on a whale shark's skin (opposite) can be understood as an evolutionary adaptation for conveying information—not intentionally, but instead passively, by blending with the patterns of sun-dappled water.

ADAPTED TO COMMUNICATE

Communication comes in modes as varied and marvelous as the animal kingdom's physical forms.

Much of it occurs outside our realm of perception. Philippine tarsiers have among the highest-pitched voices of any terrestrial mammal; they communicate in pure ultrasound, a sonic niche that may help them avoid detection by predators in the rainforests where they evolved. Likewise the black jacobin hummingbirds of Brazil, who were long thought to make no sound at all. By using equipment originally designed to record bat echolocation calls, then replaying them at dramatically slowed speeds, researchers learned that jacobins sing quite melodically.

It's not only extra-high-pitched sounds that we cannot detect. Mantis shrimp, reef-dwelling crustaceans known for snapping a pair of hinged, club-like arms with such speed and power that they cause water to boil, are the only animals known to perceive circular polarized light. Mormyrid fish, along with sharks, rays, catfish, duck-billed platypuses, and some echidnas and salamanders, even detect and generate electric signals. So-called electroreception is believed to be among the oldest sensory systems, used in everything from navigation to social interaction; it was possessed by the earliest vertebrates but lost in the evolutionary passage from aquatic to terrestrial life before evolving again.

PHYSICAL WONDERS

In mormyrids, that ancient ability functions through a specialized electric pulse-generating organ, located in their tail, and a network of nerves connected to jelly-filled, electroreceptive pores in their skin. What might it feel like to perceive electricity? Could it be something like the sensation of a breeze on one's arm hairs, only more precise?

If this is difficult to imagine, some communicative specializations are remarkable in more immediately accessible ways. The fastest muscle twitches of any vertebrate species can be found in the vocal cords of birds. (Some birds move with extraordinary rapidity in other ways, too. Viewed on high-speed video, the hops of blue-capped cordon bleu birds are actually intricate dances.)

Spring peepers, the fingertip-size frogs whose springtime wetland choruses reach decibel levels found at stadium rock concerts, achieve that volume by harnessing the most energy-efficient muscles yet found in a cold-blooded animal. They're optimized for converting caloric energy into noise.

THE COURTSHIP displays of blue-capped cordon bleu birds (above) appear to the naked eye as simple hops. When slowed down on high-speed video, though, those hops are actually intricate tap dances that produce a rhythmic accompaniment to the birds' songs.

SEBA'S SHORT-TAILED BATS (opposite), spread the seeds of fruits they eat. In turn the fruit trees have evolved to be more easily noticed. It's communication across evolutionary time.

MIXED SIGNALS

Confronted by rattlesnakes, which can detect infrared radiation, California ground squirrels dilate the blood vessels in their tails, raising the temperature of their tail to match the temperature of the rest of their body. The squirrels suddenly appear to be a larger, more threatening animal to the snakes.

Impressive as that deception is, though, it's got nothing on Alcon blue caterpillars, who secrete chemicals so similar to those of ants that ant colonies will raise them as their own queens. Unwitting ants carry the caterpillars into the colony and become their servants—but the signal-exploiting pupa are victimized in turn by parasitic wasps who lay eggs inside the chrysalis. Months later, the wasps hatch and are in turn attacked by the colony, but they emit chemicals that cause the ants to attack one another. The wasps then sneak out in the resulting confusion.

EVOLUTIONARY WINDOWS

The evolutionary back-and-forth implied by that intricate multispecies cycle, the countless generations of mutations and countermutations, is mind-boggling.

Animal communication often tells a story of adaptation across time: The elaborate coloration of diurnal butterflies, for example, is well suited to catching the attention of prospective mates during sunlit daytime flights, while nocturnal moths instead rely on chemical cues to find one another in darkness.

Some species have adapted not only to communicate with one another, but across the biological world's boundaries. Plants that rely on bats to spread their seeds have evolved leafless, pendulous branches that attract bats by reflecting their echolocation calls. The colors of fleshy fruits signal fat content to the migratory birds who eat and disperse them. Some flowers even generate electric fields that their pollinators perceive.

That such vegetative signals are not necessarily understood at a conscious level, and—notwithstanding intriguing and much debated research into plants displaying intelligent behavior (see page 109)—are not generally thought to be consciously generated, is beside the point. It's still communication. ⦿

SO MUCH animal communication occurs beyond the boundaries of our own perceptions. Philippine tarsiers (opposite) have among the highest-pitched voices of any terrestrial mammal; we can't even hear them. Their ultrasound calls serve as a private messaging channel.

POLLEN STICKS to fine, static-charged hairs covering the bodies of carpenter bees (above); the bees may even detect faint electric fields generated by flowers, giving them another way of assessing a plant's qualities. Many other insects are thought to be similarly adapted to perceive electricity.

Cues, signals, messages: Life is a constant gathering of information.

ODOR

Given the enormous importance of language to us, and the verbal origins of most languages, other possible modes of communication receive comparatively little attention. Of these, odor is perhaps the least noticed of all.

An average dog's nose contains some 300 million scent receptors, compared to just five million in our own. To our insensate noses, smell isn't often received as a signal. It's just something that happens. Yet odorants, or free-floating compounds that trigger cellular receptors, were perhaps the original form of communication, predating the specialized olfactory systems that evolved to better perceive them. They remain a powerfully widespread and efficient medium.

Some of the most fantastic examples come from the invertebrate world. Moths locate and attract mates using pheromones; in a reversal of the animal kingdom's usual roles, females do the "calling" and males seek them out. There are more than 130,000 moth species, of which the pheromones of just 2,000 have been described—a tiny fraction, but enough to fascinate biologists with their precise, complex forms and their effects on moth neurocircuitry. The right smells go straight to their brains.

The same might be said of male dogs who catch a whiff of an ovulating female, of course. It's not just bugs that rely on odor.

For many species, scents are individually distinctive and serve as

CASE STUDY

»Dogs See With Their Noses

Mirrors are sometimes used as a scientific test of self-awareness: Animals who recognize and inspect their own reflection are said to have a humanlike sense of themselves as selves, distinct from others and the world. Dogs, like most species, don't pass this test—but they do recognize their own odor.

Scent, rather than sight, provides a better gauge of dogs' selfness, and little wonder. Their exquisitely sensitive noses perceive fragrances hidden to our own; when dogs on a walk stop to sniff, again and again and again, they're reading an olfactory story. And just as dogs are attuned to subtle cues in their human companions' expressions and gestures, so too can they differentiate between the sweat of people who are fearful and those who are happy.

THOUGH RELATIVELY insignificant to us as a means of communication, odor is vitally important to many species. Moose (above) leave messages for their neighbors by rubbing against communal trees.

markers of identity. Dogs know who they're smelling; zebrafish distinguish their kin from unrelated individuals by scent. And while some chemically mediated interactions are relatively simple—male Bosca's newts expend less effort on courtship displays when they smell other newts nearby, perhaps because competition isn't worth the effort—others are quite sophisticated.

Many mammals actually use scent to exchange information. Beavers deposit scents to mark their territory, as do pouched rats and house mice. Unlike other types of signals, which can only be perceived in the moment, these can be encountered hours or days after they're made. Odor makes communication possible across time. And even when intended for a particular recipient, scent messages can be accessed by many. (One study of ocelot latrines found that 14 other mammal species visited them, adding their own scents and investigating existing ones.) They can be understood as a form of network communication.

Moose and American bison rub their scents on trees, as do bears, who possess scent glands in their feet and adopt a specialized gait

to leave marks on the landscape. White rhinos deposit dung in communal middens, or dung heaps. Such sites function as centers of information exchange; they're not merely territorial marking posts, but places where individuals can also learn about the sex, age, status, and sexual circumstances of their neighbors.

The rhinos' deposition of dung is no willy-nilly affair. Instead, it's carefully calibrated, with males even choosing where in the midden to poop and how much they should leave behind. Scatological amusements aside, those habits emphasize their role in rhino society. The dung heaps are odorous bulletin boards.

ALTHOUGH BUTTERFLIES rely on colorful wing patterns to catch the attention of prospective mates, luna moths (opposite) and others of their kind use chemical cues. This allows them to find one another under cover of night.

BEAVERS (right) exchange social information at so-called scent mounds—piles of mud erected at the edge of their territory and marked with scents. These let passersby know that other beavers are present and provide some basic information about their identity. Sockeye salmon (above) recognize their siblings by scent.

COLORATION

Most of the animal world's basic forms first appeared about 540 million years ago, during what's known as the Cambrian Explosion. Within a few million years—a seemingly long time, but an eyeblink in Earth's history—the first bilateral body plans evolved, as well as the first spinal cords and also the first known eyes.

Some scientists think the emergence of vision helped trigger the Cambrian's evolutionary efflorescence. Sparse fossil records make it challenging to confirm that proposition, but the very possibility spotlights the importance of sight to life's trajectories. With a sensitivity to light and dark and wavelength came visual signals and meaningful coloration.

Camouflage, the ability to blend into one's surroundings, is a form of communication. Geckos whose skin resembles a dry leaf convey visual misinformation to a predator's gaze. Predation is an especially powerful driver of natural selection on coloration; so too are reproduction and the choice of mates, both often influenced by appearance.

Although it's quite possible that female black-tailed godwits, members of the sandpiper family, take aesthetic pleasure in the orange breasts and dappled feathers of males, their pigments are also linked to immune system function.

AS WITH many birds, black-tailed godwits (opposite) become more colorful during breeding season. The melanin in their feathers' pigments also interacts with their metabolic and immune systems, making their plumage both exquisitely elegant and a visual signifier of health.

CEPHALOPODS LIKE common cuttlefish (below) are renowned for the camouflaging properties of their color-changing skin. Some can also use their skin to send visual messages; the patterns may even follow grammatical rules.

Colors can be a sign of health—for example, the bright yellow stomachs of parasite-free common wall lizards, or the striking hues of parrots whose feather pigments are found nowhere else in nature and may have antioxidant properties.

Even the appearance of fruit flies involves exoskeleton compounds intertwined with the molecular activity of cellular pathways that affect longevity and reproduction. Beauty is more than skin-deep.

Coloration is not an intentional message—a male peacock doesn't decide upon the blues and greens of his splendid tail; passion vine butterflies don't know that their wing patterns warn hungry birds that they're poisonous—but it can be incorporated into other actions, such as courtship displays, that use color in a directed way.

Some creatures use light itself. Bioluminescent invertebrates, such as tiny crustaceans, sometimes called "firefleas," and the family of beetles known and loved as fireflies, attract mates and disorient predators with illumination generated by chemical reactions inside their bodies.

Though a firefly's glow lasts but a fraction of a second, other visual signals are more durable. In humans, writing is a form of stored, visually mediated information; this is rare among other animals, but hints of it are seen in bird nest decorations. Black kites adorn their nests with white plastic strips that signify the big raptors' strength and territorial quality. Rock sparrows outfit their own nests with feathers from other birds. Originally thought to help insulate eggs, the ornaments instead seem to signal the status of females.

Some of the best known examples of coloration come from cephalopods, many of whom have special pigment cells that allow them to change the colors

and patterns of their tentacled bodies. Although their feats of camouflage, literally seeming to vanish as they adopt seafloor textures, are renowned, many cephalopods also use their skin to communicate with their own kind. Oval squid interactions involve complex skin signals with language-like properties, including sequences of chromatic "words" that possess multiple meanings. They broadcast their thoughts on their bodies.

What could they be talking about? Cephalopods' lineage diverged from the rest of the animal kingdom some 600 million years ago, and most of their neurons are found in their tentacles. Some scientists speculate that cephalopods—both extremely smart and physiologically unique—are the closest thing on Earth to "alien" intelligence.

Evolution's ability to fashion similar outcomes with different tools, as with fish who feel pain despite lacking the neurological structures responsible for pain in mammals, shouldn't be overlooked. However, it's fun to imagine that cephalopod thoughts could be as unique as their manner of expression.

MARINE CRUSTACEANS sometimes called "firefleas" (above) luminesce en masse in a long-lasting cloud that confuses predators, or in precise pulses generated by males to attract female attention.

MANTIS SHRIMP (opposite) are the only animals known to perceive circular polarized light. Their eyes also contain 12 or more color receptors; human eyes contain three.

GESTURE

Relying as we do on verbal communication—even our visual forms of information, such as writing, represent what can be spoken aloud—it's easy to forget how informative physical movements can be.

When next you talk, pay attention to your hands, which likely move without your being consciously aware of them. Although not every such movement expresses a meaning or intention, which is required to meet the definition of gesture, many do: a downward chop of emphasis, an upturned palm of exasperation.

These motions underscore just how intertwined thought and movement are. It's even been proposed that human languages emerged from the gestures of our distant ancestors.

Gestures are particularly well studied in other primates. Great apes typically have repertoires containing dozens of signs; they can be quite versatile, with one movement having multiple meanings depending on circumstance. Many gestures human toddlers use are also used by chimpanzees, our closest living relative. One study found that of 52 gestures observed

SO MUCH can be conveyed by movement alone. A foraging honeybee (opposite) performs an intricate dance that informs hivemates about the location of nearby pollen-bearing flowers.

GREAT MINDS

» Karl von Frisch (1886-1982)

When honeybee scouts return to the hive after finding a patch of flowers, they perform an intricate dance that, by means of abdomen wagging and wing movements and the path of their footsteps, conveys to hive members the quality and location of nectar.

In the mid-1920s, when biologist Karl von Frisch started to decode these dances, most scientists were skeptical. They believed that bees simply followed their antennae. The very thought of them possessing something like language—and then expressing abstracted information through dance—was radical, if not downright preposterous.

A half century later, Von Frisch received a Nobel Prize for his research. Other researchers have added new details: Honeybees also dance to communicate about new nest sites and to ask nestmates for help grooming body parts they cannot reach on their own.

in one- to two-year-old children in Germany and Uganda, 46 are also used by chimpanzees. It seems we have an innate repertoire common to our lineage.

Gestures have not only fruited upon our branch of evolution's tree, and gestures don't always require hands and arms. Lionfish invite each other to join group hunts by flaring their fanlike fins. Magpies, though known for their especially rich, complex vocalizations, also use their beaks to point at objects. Big, bushy gray squirrel tails, so helpful in maintaining balance on swaying tree limbs, are also quite expressive: Squirrels use them to signal aggression and warn other squirrels about approaching predators. Tail flicks can also signify agitation or frustration.

For squirrels, tail gestures convey information quietly, without drawing attention to themselves. Cow gestures—the position of their ears, the movement of their tails—are likewise shaped by a need, honed by millions of years of predator gazes, for subtlety. The practice of tail docking, in which a dairy cow's tail is amputated, is controversial in part because it removes one of her methods of communication.

Gestures are also quite evident in what's known as rough-and-tumble play. Found not just in rambunctious kids but also in many other mammals, such play is invaluable for developing physical coordination and social grace. There's always a danger, however, of roughhousing escalating into full-blown fighting, so good communication is essential—and, from rats and dogs to monkeys and apes, gestures are an important part of the back-and-forth that keeps their rough-and-tumble playful.

These dynamics may have helped to shape the evolution of gesture and its cognitive underpinnings. From that perspective, language didn't just arise from gesture, but out of playfulness.

AUSTRALIAN MAGPIES (above) are vocally prolific animals, with vocabularies containing many hundreds of syllables—but simply using their beaks to point still comes in handy, and underscores how hands are not required to gesture. Most any purposeful body movement will do.

ACCORDING TO some theories, modern human languages have evolutionary roots in our ancestors' gestures. The gestural repertoires of gorillas (opposite) contain dozens of signs; some are inborn but others are learned, and may even be unique to a group or an individual.

We rely so much on speech that it's easy to forget how informative physical movements can be.

TOUCH

Though quite a bit of information can be conveyed in a handshake or a hug, touch is a comparatively minor part of our own communicative life. For some animals, though, touch is central to information exchange. It is even performed, with the use of vibration, at a distance.

Many spiders, for example, use their webs not only to capture prey but to send and receive messages plucked upon the strands, even using specialized threads designed to conduct these signals. Others, such as male brush-legged wolf spiders, capture female attention by tapping their forelegs on leaf litter. This allows them to transmit more efficiently than they could by making noise, and particularly intense rhythms signify the suitors' robustness.

No fewer than 150,000 insect species communicate with vibrations transmitted along surfaces and through objects. Among them are leafhoppers and planthoppers, families of insects whose messages ripple through the leaves and stems on which they live. Some have even evolved the frequency of their calls to resonate with their host plant species. Their plants can be understood as transmission channels, and they discern one another's "voices" from movements caused by wind and rain.

Some termites signal alarm by drumming their heads on the

GREAT MINDS

» Katy Payne (b. 1937)

In 1984, biologist Katy Payne spent a week observing elephants at Oregon's Washington Park Zoo. In their presence she felt deep, subtle vibrations, something like the lowest notes of an enormous church organ.

Payne traveled to Africa and together with Joyce Poole described how the elephants' rumbles reverberate for miles through soil—at a frequency beneath the limit of our own hearing—and are detected by vibration-amplifying pads of fat in distant elephants' strong, sensitive feet.

This research might help people protect elephants. It's certainly informed our sense of what it's like to be an elephant, constantly connected at even great distances to one's family and friends.

ground. As a signal starts to fade, it's relayed by another head-drummer, traveling through the colony at speeds of up to several feet a second. Fiddler crab foot drumming is transmitted through sand and amplified by vibration receptors in the recipients' legs, which act as biological signal boosters.

Vibration is less common but still widespread among vertebrates. Several amphibian species practice so-called seismic communication; by tapping their feet, white-lipped tree frogs bypass the sonic din caused by their vocalizations, and the resulting signals travel through a neighbor's legs and directly into their inner ears. Mimicking rain or digging predators, wood turtles lure subterranean earthworms from their burrows and up to the earth's surface where they become an easy meal for the hungry reptiles. Touch, like any other form of communication, can be deceiving.

Male gray seals sizing each other up for a fight will slap their bodies on the ground. The resulting shock waves are reliable indicators of body size, giving smaller individuals an opportunity to back down. Most vibration-using mammals, however, such as mole rats and kangaroo rats, live underground

BOTH ASIAN and African elephants (above) produce deep rumbles that travel for miles through the ground and are detected by other elephants' feet—but a trunk's intimate touch is marvelous, too.

SEISMIC communication doesn't need to be long-distance. Sending information via vibrations across just a few inches of soil allows eastern newts (opposite) to exchange messages without drawing attention.

LEAFHOPPERS (above) and planthoppers transmit vibrations through the vegetation on which they live. Some species have coevolved with their host plants; those stems and leaves resonate with the frequency of their calls.

KANGAROO RATS (right)—technically not rats but relatives of pocket gophers—dig burrows in desert soil. Safely ensconced, they use their feet to drum out messages to their subterranean neighbors.

and bang their feet and tails and heads against tunnel walls. With these signals they attract mates, warn one another about predators, and establish territories. They could be saying much more, of course; scientists have only started to translate them.

Unlike in arthropods, for whom vibratory communication is an ancient trait, this is a relatively recent adaptation for mammals. It originated some 45 million years ago, in tandem with a period of global cooling that drove their ancestors underground. Unable to otherwise see or hear one another, they learned to send messages through the earth itself.

SOUND

From the telegraph to the internet, modern telecommunications have captured human imagination by making it possible to transmit large quantities of information across long distances at near-instantaneous speed. In a way, these technologies are just expanding on the essence of sound itself.

Sonic communication is ubiquitous in the living world. Often it takes surprising forms. Unlike moths who rely on chemical signals, female Australian whistling moths sing to prospective mates, as do female short-bodied cellar spiders, who during copulation use song to influence the movements of their partners.

In considering the manifold marvels of birdsong, it's hard to know where to start: perhaps not with any single species, but rather with their collective voices at dawn. Shortly before sunrise, songbirds sing in unison at intensities not heard at other times of day. Many scientific explanations have been offered for the chorus, from establishing territory to attracting mates, but no single explanation yet suffices. There are likely many that vary between species.

As for why the choruses occur precisely at dawn, it's thought to be linked to atmospheric dynamics. As the sun rises, its rays heat landscapes unevenly; different types of vegetation absorb differing

BLACK-CAPPED DONACOBIUS (opposite) were once thought to be members of Mimidae, the especially vocal family to which mockingbirds and catbirds belong. That taxonomic designation proved incorrect but understandable—black-capped donacobius are a taxonomic family unto themselves. The duets performed by mated pairs are loud and exceptionally complicated.

THOUGH MOST moths depend on chemical signals to attract and detect mates, whistling moths (below) are among the few species to make sound. They "whistle" by rubbing their forelegs together.

amounts of radiation, and rolling terrain likewise varies in temperature, producing a thermal patchwork of subtly turbulent air. Dawn's stillness is when songs carry most clearly.

Many birds don't sing their own songs but mimic those of others. Northern mockingbirds perform glorious extended mash-ups containing the voices of more than 100 species. And some birds don't just sing, but also make noise with their feathers. Long thought to chirp with their mouths, Anna's hummingbirds actually chirp with their tails.

In a resonant mammal example, porcupine-like tenrecs make noise by rubbing their quills together. Sound generated by the rubbing of body parts is called stridulation. It's known mostly from the insect world, where grasshoppers stroke their legs upon their wings, like bows across a violin's strings, and crickets rasp together their wings. Mole crickets even fashion their own concert halls, digging burrows that amplify their calls.

> The essence of sonic communication is sending lots of information across long distances at very high speeds.

The aquatic world is as noisy as the terrestrial. Humpback whale songs pass from clan to clan like so many popular hits; blue whale songs literally cross oceans. And though people don't tend to think of fish as communicative, many use their swim bladders—the gas-filled sacs that help fish maintain buoyancy—as drums. The bladders come in many forms, as do sonically adapted fins and spines. Fish actually possess a greater diversity of sound-generating organs than any other vertebrate clade.

As with birds, some fish have their own dawn choruses. When Gulf corvina gather each year to spawn in the Colorado River Delta, their combined voices yield some of the loudest sounds ever detected in marine environments, reaching peaks above 200 decibels—considerably louder than even stadium rock concerts, which typically max out around 120 decibels. Unfortunately, Gulf corvina are threatened by overfishing. What's at risk isn't just the species, but one of nature's most incredible shows. ⦿

THE CHIRPS of Anna's hummingbirds (above left) are generated by wind rushing across their tail feathers as they dive. Many birds make noise with their feathers; subtle changes to their shape fine-tune the sounds.

STRIDULATION, or the production of sound by rubbing body parts together, is best known from crickets and grasshoppers. Some mammals do it too, including tenrecs (above right) who stridulate with their quills.

GREAT MINDS

» Roger Payne (b. 1935)

In the late 1960s, biologist Roger Payne and his wife, Katy, visited a Navy engineer who played them underwater recordings of humpback whales. The cetaceans were not known to sing. The engineer, worried that his discovery might be used by hunters, told only a few people. The Paynes pioneered the study of these songs, which are part of the whales' culture. Performed by groups of males, songs can last up to 30 minutes, weaving themes and motifs, and changing subtly by season. In 1970, Roger released *Songs of the Humpback Whale,* a collection of recordings that would become the best-selling nature record ever.

CHAOTIC VOICES

Physicists have a term for the dynamics of systems that, like an avalanche-prone slope of snow or a pot of water just before it hits the boiling point, are poised at the edge of chaos: They exhibit "nonlinearities," tiny fluctuations—like an ill-fated footstep or one more degree Fahrenheit—that can produce massive changes. The subsequent chain of reactions is so complex that it defies straightforward prediction.

There's no more urgent noise than a baby's cry—and in it is the sound of chaos.

Nonlinearities can also be found in the voices of animals. Fabulously convoluted birdsongs result from nonlinearities generated as air flows through birds' vocal organs. Called syrinxes, the nonlinear effects these organs produce can be compared to a distortion pedal that warps an electric guitar's signal. (Distortion is one kind of nonlinearity. Jimi Hendrix was a pioneer of acoustic chaos.)

Most frogs have relatively simple vocalizations, but concave-eared torrent frogs, equipped with nonlinearity-generating larynxes, sing with birdlike richness. Toadfish, whose

SO-CALLED nonlinear noises contain an element of sonic unpredictability that makes them neurologically attention getting. This is advantageous when calling in distress or alarm—or, in the case of female koalas (above), when rejecting the attentions of unwanted males.

THE VOCAL gymnastics of birdsong are biomechanically simple. The physical properties of their vocal organs allow species like black-capped vireos (opposite) to produce fabulously complex sounds with minimal effort.

humming "voices" are generated by rapidly twitching the muscles of their swim bladder, likewise harness nonlinearities to add nuance to an otherwise simple sound.

Among the different types of nonlinear noise are biphonation, or the simultaneous generation of two pitches; subharmonics produced by the interaction of different sonic frequencies; and rapid fluctuations in frequency. These are complicated terms, but the principles guiding their evolution in animal communication are straightforward. Nonlinearities yield complex signals with minimal effort.

They're also attention grabbing. According to one theory, nonlinear noises contain, from millisecond to millisecond, an element of unpredictability, and compared to a monotonously repetitive signal, they're hard to tune out. This could explain why nonlinearities are so often present in mammal alarm calls. Dog barks are nonlinear, as are the sounds female koalas make when rejecting an unwanted male's advances—this being, for them, a matter of considerable import, as those noises let other males know they're still available.

Other examples of nonlinear vocalization are the sound of our own cries and the wailing of babies. There's no more urgent noise, and in it is the sound of chaos.

SURPRISING SOPHISTICATION

Evolution excels at turning a few simple components into systems of considerable sophistication. Nonlinear noise (see pages 34–35) is one such trick. Another is multimodal communication, or the combination of more than one kind of signal at a time.

An everyday example comes from gray squirrels. By combining vocal alarm calls with tail waving, they can specify whether a predator approaches by land or air. Multimodality also helps when one transmission channel is jammed: Bornean rock frogs whose calls are drowned out by the roar of fast-moving water communicate by waving their feet and arms.

Using several expressions simultaneously also enhances meaning, as when a grimace of pain heightens the urgency of a distress cry. Communication also can be limited to a single channel, even a single sound, but with new meanings made possible in clever ways. By repeating one sound at varying rates, pied babblers produce "purrs" and "clucks" that, respectively, call their young toward food or signal a desire to move elsewhere.

If an animal has the requisite cognitive sophistication, signals can also change their meanings according to context. Among humans, a waving hand can mean both hello and goodbye, delivered by a stranded motorist it's a call for help, and in a restaurant it catches a waiter's eye.

Context is vitally important for many other animals, too. Much of baboon interaction is mediated by grunts that, on their own, don't seem to provide much information—but, exchanged by longtime friends after a fight, grunts signify reconciliation.

One study of chimpanzees found that a gesture signifying "touch" had 36 different forms used in 26 separate contexts. A single "hoo" call, which conveys a desire for group members to stay close, can—depending on whether chimps are resting, traveling, or alerted to a threat—instruct them to stay put, approach the caller, or approach slowly.

Delivered by a mother bear to her gamboling cubs, a guttural moan of displeasure can mean "Get out of the tree!" A cub might use the same sound in expressing a desire to be left alone. Even with limited means, much can be said. ⦿

UNLIKE THEIR savanna-dwelling baboon cousins who express reconciliation with grunts, mandrills (opposite) reconcile by silently baring their teeth. It's a simple gesture, yet depending on context and the relationships of the individuals involved, can be quite meaningful.

COURTSHIP DISPLAYS of wolf spiders (below) involve leg waving and foot tapping. They're studied by scientists interested in the fundamental principles of communications that combine multiple kinds of signals.

ELEMENTS OF LANGUAGE

No other species has a system of communication that's exactly like human language, but many basic properties of language can be found throughout the animal kingdom.

Scientists once thought animals could only express themselves in involuntary, emotional ways: a cry of pain, a grunt of satisfaction. Communicating intentionally about something other than one's own immediate state of mind—so-called referential communication—was considered uniquely human.

This is certainly not the case. A coyote pup who "play bows" toward a sibling or a dog who drops a ball at your feet is demonstrating intent. One could also think of those gestures as referring to play.

Referential communication has been rigorously described in many species. Landmark studies involved monkeys whose alarm calls referred to different types of predators. As with so much research on animal intelligence, traits first observed in our closest relatives were soon described in other, less-related creatures.

The first birds demonstrated to use referential communication were domestic chickens whose calls alerted flock members to the location of food. Noisy miners, a bird species native to Australia, use alarms that—like the

GREAT MINDS

» Toshitaka Suzuki (b. 1983)

To determine whether small songbirds called Japanese tits use syntax, biologist Toshitaka Suzuki recorded their calls and then replayed the sounds for them. One set of warning calls prompted the birds to look around and approach the loudspeaker—but played in reverse, it occasioned no response. Order mattered. The birds used syntax.

In another experiment, Suzuki broadcast the birds' snake-specific and general alarm calls, then observed their reactions to a stick that he jiggled like a snake. Only after hearing the snake call did the birds seem to look for a snake, suggesting that specific sounds are accompanied by images in their minds' eyes.

SCIENTISTS once considered animals incapable of communicating intentionally, with reference to objects or events rather than emotional states. Among the species to dispel this notion were domestic chickens (above).

monkeys—distinguish between raptors approaching by air or land. Coral trout, who partner with moray eels to hunt, gesture with their heads to guide the eels toward prey hiding in deep rock crevasses.

Prairie dog communication is especially rich. Their high-pitched squeaks can inform colony members not only of an approaching human but of the color of the intruder's shirt, whether they're tall or short, and the speed of their approach.

One especially important element of language is syntax, or the ability to produce new meanings by arranging words in different order. ("Open the door" and "the open door" indicate two different things.) Syntax has been described in the calls of Japanese tits, small songbirds related to chickadees (see sidebar opposite), and in Carolina chickadees as well.

These birds produce many fewer sounds than we do, but simple syntax helps make the most of what they have. And although more complex forms of grammar—such as embedding messages within other messages, as demonstrated in this sentence—are

exceptionally developed in human language, European starling calls also appear to contain nested grammatical forms.

Human languages are also characterized by turn-taking: the back-and-forth alternating patterns characteristic of conversation, hints of which can already be heard in the babbling of infants who haven't yet learned to talk. Scientists have studied turn-taking in no fewer than 360 bird species, most primates, cetaceans, bats, rodents, four orders of insects, and frogs and toads.

Their turn-taking isn't necessarily conversational. When male frogs alternate their ribbits, they're not talking to each other so much as they're avoiding the wasted effort of trying to call over one another. Nevertheless, in that interaction the echoes of our own voices can be heard.

PRAIRIE DOGS (opposite) have communication systems so rich that they've been compared to human language. The scientists who study them hope this insight will make people more sympathetic to the rodents, who are often treated as vermin.

IT'S NOT ONLY overtly brainy animals like primates and elephants who have language-like communication systems. Japanese tit songbirds (right) arrange their calls syntactically: Meaning changes with sound order. Coral trout (below) were among the first fish to demonstrate referential communication.

HUMAN LANGUAGE

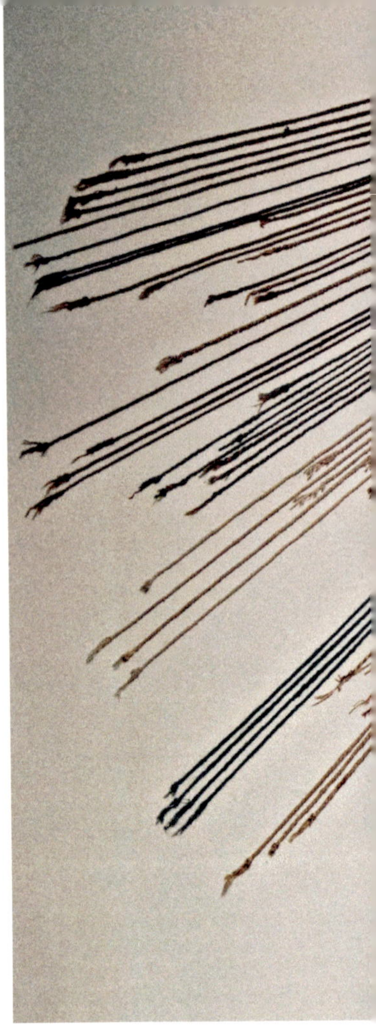

Much remains to be learned about human language, including the fundamental question of when and how it evolved.

Was this a gradual process, with linguistic faculties accreting piece by piece across evolutionary time, so that one could draw a line between the properties identified in other animals—gesture and facial expressions, inference and reference—to their contemporary flowering in *Homo sapiens*? Or does human language represent a difference not just of degree but also of kind, evolving not in stepwise fashion but instead taking a massive, sudden leap in us? And to what extent have human anatomy and social structures shaped it?

These questions have no consensus answer. (It would probably help too if our closest ancestors were still alive. Studies of other great apes are quite informative, but what might we learn if we could listen to *Homo erectus,* the species that gave rise to us?)

That said, language is certainly exceptional—and although its differences with animal communications are sometimes overstated, and its absence in animals is too often seen as evidence of their limitations, that shouldn't detract from our appreciation of just how wonderful language is.

GREAT MINDS

» Mark Pagel (b. 1954)

More than a decade ago, evolutionary biologist Mark Pagel suggested that evolution didn't always proceed steadily but sometimes erupted in bursts of change. He wasn't the first to say so—but Pagel also studied language, and languages too, he concluded, evolved along the same jumpy trajectories as species. To Pagel, the evolution of language is a window into the evolution of culture. And although other animals also have cultures, our own are—thanks in part to language—uniquely complex, transmitting enormous quantities of information and enabling rapid social change. Human cultural evolution may now be as fundamental a force as biological evolution.

Grammar makes possible constructs of staggering complexity: The longest known English sentence contains 13,955 words, or almost as many words as you're now holding in your hands. Human languages are uniquely able to represent abstract concepts, and uniquely versatile in coining new words and phrases. They also go meta quickly. To comprehend the sentence "Peter thinks that Mary said that it is implausible that pigs fly" is as singular and important a trait, say some scientists, as echolocation is for bats.

Language comes in many varieties—Romance languages, in which a word's meaning resides in its unique combination of consonants and vowels, and tonal languages, in which intonation can give one word radically different meanings—and forms. Languages can be whistled and gestured; they can be represented by ink on a page, or even knots in string.

QUIPU, or "talking knots" (above), were used by many precolonial cultures in the Andes Mountains of western South America. Combinations of knots were used to keep records, such as inventories of crops, and perhaps to encode histories and biographies. Modern archaeologists have yet to fully decipher them.

Much remains to be learned about human language, including the fundamental question of when and how it evolved.

PART TWO

SOCIAL TALK

Most animals live in groups, and their social worlds are composed of relationships founded on communication.

Chemical cues enable juvenile salmon to recognize their siblings, and thus to treat them differently from unrelated salmon. Red-crowned cranes perform intricate dances, perhaps as expressions of devotion, both during courtship and throughout their monogamous unions.

Some species—including African and Asian elephants, bonobos, American crows, and humans—live in "fission-fusion societies," which regularly break into smaller groups and later reunite. This lifestyle shapes especially rich communicative capacities that sustain complex social systems across distance and time.

When two individuals meet after a prolonged absence, or for the first time, how can they negotiate the uncertainty inherent in that encounter? It helps to read body language and physical characteristics—the size of a side-blotched lizard's tail signifies social dominance—and exchange greetings.

Long-term relationships would be difficult to maintain without communication. So would complex societies and communal decisions, which other animals sometimes make democratically: African wild dogs "vote" on pack movements by sneezing. Even seemingly simple animals can accomplish extraordinary things by exchanging information. Ant colonies can be so large and complex that they're likened to human societies; it's all organized with chemical signals.

The study of animal communication isn't just about how animals exchange messages. It's about what these messages make possible.

THE DIVERSITY of biological communication systems isn't just a wonderful chapter of natural history. It's the foundation of social life. For mated red-crowned cranes (opposite), ritual dances help nourish a lifelong relationship.

EMOTION

For most of the past century, scientists tended to regard animal emotions as quite simple. Animal communication was similarly dismissed as involuntary, capable only of expressing rudimentary internal states.

To modern sensibilities, this can seem quite strange, not least because certain involuntary expressions of feeling—whoops of laughter, cries of joy or distress—are such important parts of our own lives. Yet some scientists even considered animals to be unconscious of their own feelings. After all, they lacked the language supposedly necessary to transmute emotional impulses into meaningful experiences.

Nowadays, the absence of human-style language is not considered a barrier to intelligence. Scientists have also started to recognize that animal emotions have much in common with our own. And although animals can communicate about more than emotion, expressions of feeling are essential parts of their inventory.

Perhaps the most universal of these expressions are the cries of infants, which are not only biologically difficult to ignore but also have common acoustic properties shared by other mammals, birds, reptiles, and amphibians. Some animal mothers even respond to sounds of upset babies belonging to other species.

Adult distress can also transcend species boundaries, as in 2013, when the moans of dairy cows whose calves had been taken away unsettled the human residents of a Massachusetts town. Whether such cries are considered involuntary makes them no less poignant. Cows are instructive in another way, too: Having evolved under the watchful gaze of predators, they often convey emotions subtly, with only tail flicks or the position of their ears betraying their feelings.

Expressions of well-being are also shared across distantly related creatures. Rats laugh while playing and when they're tickled; their voices are so high-pitched, though, that we cannot hear them. Kea parrots also produce a laughter-like call while playing, either alone or with other parrots. Researchers suspect that it lifts solitary spirits and, when made in a group setting, functions like contagious laughter does in people, spreading good feelings among company.

ANIMAL communication was long said to be unsophisticated. Many scientists also considered animal emotions—like the care of a mother cow for her calf (opposite)—to have little in common with our own. We now know that to be untrue.

WHEN SOMEONE laughs, people nearby are quick to join. This so-called contagious laughter is a basic human trait that helps reinforce social bonds among large groups; something similar has also been described in kea parrots (below).

ALARM

Of all the meanings that can be expressed, perhaps none have yielded more insight into animal communication than alarm.

Calls of alarm, voiced by many social creatures in response to approaching predators, are well suited to experimental investigation. Researchers can record the calls, play them back, and observe how animals react. Given the urgency of the alarms and the clearly defined responses—Run! Hide!—it's not hard to figure out what they mean.

Scientists have paid particular attention to alarms made by primates, from our closest great ape relatives to more distant branches of the family tree. A landmark 1980 study involved vervet monkeys: They use distinctly different alarms for leopards, eagles, and snakes, and researchers found that each call evoked a corresponding behavior. Leopard alarms sent monkeys scurrying up a tree to safety. Eagle alerts pulled their gazes skyward. News of a snake caused them to look down.

Several basic principles could be inferred from their reactions. The alarm calls contained specific information about each predator. Other monkeys understood that information. It sounds quite simple, even self-evident, but it was paradigm shifting.

Until then, scientific convention held that animal communications were limited to emotional outbursts. The ability to refer to

GREAT MINDS

» Dorothy Cheney (1950–2018) and Robert Seyfarth (b. 1948)

In the early 1970s, Dorothy Cheney and Robert Seyfarth became research partners as well as spouses. They would go on to transform the study of animal communication, uniting the methodological rigor of lab-based experiments with a passion for studying wild animals. Their research on vervet monkey alarms (see above) challenged conventional scientific wisdom about animal communication as strictly emotional and set an example for deciphering it.

Later, they investigated further questions: Which aspects of communication are innate, and which are learned? What do primates know about their peers? Equipped with a deeper understanding of what primates say, they illuminated their rich social worlds.

objects and events was considered beyond them. The vervets' alarms clearly showed otherwise. That primates should communicate so thoughtfully might come as no surprise—after all, they're our cousins—but further research would describe alarms in other mammals as well, particularly rodents.

Gray squirrels, like monkeys, have different alarms for aerial and terrestrial predators. Prairie dog alarms are extraordinarily detailed (see page 39). California ground squirrels use their rattlesnake-signifying alerts, which involve tail waving rather than vocalizations, to deceive: They wave their tails even when no rattlesnakes are visible, ostensibly to convince unseen snakes that the squirrels are aware of them. Meerkats—not rodents but members of the mongoose family—have specific calls not only for their predators but also to indicate the presence of their predators' poop.

Less alarm research has been conducted on animals other than mammals, but a handful of bird alarms have been described. Steller's jays make different calls for different raptors: Whether an approaching bird is a sharp-shinned hawk or a northern goshawk has life-or-death implications for how they respond. Noisy miners distinguish between

ALARM CALLS are a favorite subject of scientists studying animal communication. Meerkats (above) have calls that refer to different types of predators, and even to indicate detection of their predators' poop.

VERVET MONKEYS (above) were the subjects of some of the most significant studies ever conducted on animal communication. Their translated alarm calls conclusively demonstrated that animals can express themselves in language-like ways.

WHEN SCIENTISTS first described the language-like systems of primates, it was impressive but perhaps not surprising: After all, primates are closely related to us. The discovery of these abilities in species like noisy miners (opposite) shows just how widespread these systems are.

EASTERN GRAY SQUIRRELS (right) and other members of their genus provide an everyday window into animal communication. They announce alarm both vocally and by waving their tails. The combinations produce more meanings than would be possible with just one kind of signal.

hawks who are perched or flying. The latter are a mortal threat; the former can be beaten back, and miners recruit mobs to harass their temporarily stationary enemies.

One particularly important study involved domestic chickens who, when in the company of other chickens, announced the arrival of an oncoming hawk but stayed quiet if threatened while alone. That their alarm issuing depended on the presence of an audience was clear evidence of the chickens being able to control what they said, rather than communicating impulsively and involuntarily—another now punctured misconception about animals.

Much remains to be learned about alarms in creatures besides birds and mammals. It's important to remember, though, that alarms don't necessarily dominate animal life. Unlike a conversation about the weather or what happened yesterday, alarms are just easier for us to understand.

GREETINGS

Greetings are such a casual, ubiquitous part of our own lives that it's easy to overlook how complex and information-rich they are. A person's identity, state of mind, intentions, a sense of mutual trust, perhaps an assertion of status: All that is conveyed by a simple handshake or nod.

That our greetings are also so *physical*—we don't just say hello but use our hands and limbs and facial expressions—long ago caught the notice of researchers who analogized them to greeting rituals found in other animals.

When two dogs meet, they often sniff one another's rears—a strange if amusing habit to our eyes, but a rich source of scent-based details for them. Spotted hyenas likewise sniff one another's genitals. They live in hierarchical, female-dominated communities, and greetings between females serve to reinforce social position. A female will signal submission by presenting her pseudo-penis—hyenas are *very* interesting animals—erect.

Male guinea baboons take it even further, not only touching faces and embracing but also holding one another's scrotums. The longer they do this, the stronger their relationship tends to be. In mutual vulnerability resides trust.

Greetings can also reinforce a bond, as with monogamous pipefish who spend just a few minutes together each day, engaging in elaborate greeting displays. The

GREAT MINDS

» Barbara Smuts (b. 1950)

Much of what is known about baboon greetings and communication can be traced to anthropologist and ethologist Barbara Smuts. She spent more than a quarter century studying them in Kenya and Tanzania, and has described their social lives in extraordinary detail. For months at a time, she traveled with baboon groups from morning until dusk, defying the once conventional wisdom that primatologists needed to remain studiously apart from their subjects. Instead, Smuts learned to communicate with them, using body language and eye contact like they did. She gained their trust—and ultimately a richer understanding of their lives—by entering their social world on baboon terms.

more time they spend doing this, the stronger the bond.

AT A MORE fundamental level, greetings open a channel of communication and establish the possibility of a relationship. Oval butterflyfish are slow to recognize one another; even for mated pairs, this can take a little while. They greet their own mates with the same tail-up display used for strangers—which serves to buy some time, giving them a chance to remember that they're actually a couple before escalating to territorial aggression.

At the other end of the spectrum, orca pods have been filmed lining up to meet one another in an elaborate choreography that looks almost like a line dance. Elephants meeting after a long separation voice long, low rumbles, touch tusks, and even spin with excitement.

It's easy to view these moments through a lens of information exchange and relationship maintenance, and no doubt that's true. Yet elephant greetings are reminders that, aside from the adaptive functions their greetings serve, they're also expressions of joy at being reunited with friends and family.

Although most research on animal greetings has involved

GREETINGS ARE deceptively simple things. Much can be learned and conveyed in a few moments of mutual recognition—a principle that extends from humans to hermit crabs (above).

WHEN STUDYING animal greetings through an evolutionary lens, one can get caught up in macroscopic explanations: how information exchanged might provide reproductive advantages, the different strategies shaped by different life histories. Yet that shouldn't obscure the simple joy of three elephant friends reuniting (above).

MUCH REMAINS to be learned about the communications of orcas (opposite). People have observed pods engaging in formal, even stylized greetings; we can theorize about the evolutionary function of these rituals, but what they're saying is still unknown.

species who are highly charismatic or conspicuous, much remains to be learned about other animals—including those whose sociality is little appreciated. The scientific literature contains few mentions of salamander greetings, yet people who know tiger salamanders from captivity report them exchanging complex greetings with one another and even with people.

What might be happening when two hermit crabs, crossing paths on a beach, engage in a prolonged exchange of clicking and antennae waving? Someday, perhaps, we'll find out.

Greetings are such a casual part of our own lives that it's easy to overlook how important and information-rich they are.

COURTSHIP

Necessity is the mother of invention, and nothing is more necessary across evolutionary time than finding a good mate. No wonder some of the most complex, varied, and simply fantastic communications occur in the context of courtship.

To convey their amorous intentions, male American alligators arch their backs beneath the water and bellow at a frequency so low it makes the water's surface froth like a fountain. The deeper their voices, the larger—and presumably more attractive—they tend to be. Male Bornean tree-hole frogs tune their voices to optimal resonance with their eponymous homes, adjusting the pitch depending on how much water has collected in the bottom, thus giving males the best chance at being heard by prospective mates.

In most species the males try to impress the ladies, though sometimes the roles are reversed. Among croaking gouramis, a group of freshwater fish, females initiate courtship by approaching males, holding themselves vertically in the water, and generating sounds like a purr.

Some of the best studied courtships involve birdsongs, which have been likened to the sonic equivalent of a peacock's wonderfully ornate tail. Song complexity may indicate health: Birds who have had stressful, development-impairing early life experiences tend to have shorter, simpler songs than their more fortunate peers. It's not just physical health that matters, either. In long-billed hermit hummingbirds, a song's frequency suggests body size, but its consistency tracks with spatial memory. Females are looking for males with brain as well as brawn.

CHARLES DARWIN famously expressed frustration at peacock plumage: How could such fantastically beautiful, seemingly nonfunctional ornamentation evolve? The answer, he surmised, was through countless generations of peahens

THE SOCIAL lives of reptiles are surprisingly rich. One of nature's most impressive courtship displays belongs to American alligators (opposite), whose bellows make the water above them dance.

NOTHING IS more important than finding—and keeping—a good mate. Long-lived and monogamous oystercatchers (left) engage in acrobatic aerial duets that help reinforce the bonds between them.

PEACOCK PLUMAGE (opposite) is an iconic example of evolution by sexual selection: Characteristics preferred by one sex, in this case females, become magnified over time and may even become divorced from their original function. Nobody is quite sure about the function of peacock ornamentation; it may signify physical health.

preferring symmetrical, many-spotted plumes. He called this sexual selection, and it proved to be a crucial element in the theory of evolution. (Later research showed that peahens do, in fact, prefer males with more spots; as with birdsongs, plumage for peacocks signifies health.)

Darwin also described dance as a sexually selected signal that reveals an individual's quality. Some researchers have looked for evidence of this in human dance, and although the findings are hard to generalize—do women really prefer a man who dances like John Travolta in *Saturday Night Fever*?—there's an underlying truth, though it's not experienced in the technical language of science.

Those links between appearance and underlying traits are manifest across evolutionary time, but the lived experience is felt. A female sage grouse watching a male strut and fan and pop out his chest likely isn't thinking about the reproductive fitness of her descendants, but she may feel a jolt of attraction.

Courtship doesn't always end with mating. Partners may stay together for seasons, even lifetimes. The cooperation that entails requires not just day-to-day communication but also activities that reinforce relationships. When pairs of red-crowned cranes perform their elaborate, marvelously synchronized dances, or when oystercatchers circle their territories in acrobatic duets, they're strengthening the ties that bind them.

CASE STUDY

»Chemically Mediated Courtship

Flashy displays and beautiful songs get all the attention, but courtship takes place at the chemical level, too. Fruit flies are attracted to pheromones coded by genes involved in growth and development; simple as they are, they're still playing the field. In moths, pheromones are so complex and sophisticated that they could be likened to chemical birdsong.

Attraction between blue crabs is also triggered by pheromones: A female's chemical signals trigger a male's attention. He in turn releases his own chemicals, and expresses his intentions by standing upright, spreading his claws, and fanning out his legs. It will be several more days before the female is ready to mate. The male must carry and protect her until this time. Even crustaceans have chivalry.

COMPETITION AND STATUS

The prominence of competition and conflict in animal life is sometimes overstated. Visions of nature as "red in tooth and claw," a term first popularized in the mid-19th century, perhaps speak to our own biases, too—and a hair-raising chase makes for better wildlife documentary drama than animals just hanging out.

The *possibility* of conflict is ever present, though, and even a minor injury—an open cut, a broken tooth—can become permanently debilitating. It's far better for all concerned to settle potential disputes without resorting to blows.

Animal voices and features often convey their physical characteristics. Red deer roars bespeak their size and strength; the intense ultraviolet throats of male Augrabies flat lizards advertise their fighting ability. Duly warned, smaller and less truculent combatants have a chance to back down before anyone gets hurt.

Especially vibrant feathers also signal dominance for white-crowned sparrows, yet rock sparrows accomplish this by decorating their nests with feathers belonging to other birds. This behavior likely originated to help insulate their

MALE ELEPHANT SEALS (above) can easily weigh several tons. Their battles are commensurately ferocious. Most disputes are settled without coming to blows, though, with males instead posturing and trumpeting and slapping the ground to convey their strength.

THE POSSIBILITY of conflict is an unavoidable part of life, but conflicts themselves can be averted. Ultraviolet markings on Augrabies flat lizards (opposite) signify an aggressive temperament, giving other lizards the chance to make an informed decision about whether they really want to fight.

eggs, then was co-opted as a ritualized form of conflict mediation.

The sparrows' ornaments straddle the line between straightforward information about size or strength and more nuanced signals of social status. In a hyena's laugh can be heard her rank in the pack; the same goes for rock hyraxes, a rabbit-like mammal closely related to elephants.

Conflict can also be negotiated in more explicitly communicative ways. When tempers flare in a spirited wrestling match between young rats, they quickly talk themselves down. When fights do break out, in rats and in many other species, they're followed by gestures of conciliation. In other words—or gestures—they communicate something like, "I'm sorry."

This is vital for animals who have ongoing relationships, particularly if they're long-lived. Great apes and many other primates have a range of conciliatory expressions. Life in the animal kingdom, particularly for social creatures, is as much about apologizing and trying to set things right as it is about fighting.

PARENTING

It's a wonderful sound, a baby's babbling: the vocal patterns that form as their first syllables take shape and an innate verbal capacity gradually acquires the rhythms of spoken language.

Baby pygmy marmosets, the world's smallest monkey, also babble. So do baby sperm whales whose undirected infant clicking starts to resemble the complex, even language-like communications of adults. These communications vary from clan to clan; a young sperm whale will learn to speak in the dialect of her own family.

This process is shaped by listening to other whales, particularly their mothers, making it an especially lovely example of the formative influence—not just in whales, but also in marmosets and humans and every other mammal—of a mother's care. Well before they've learned to speak, the vocal back-and-forth between mother and infant cements their bond and calibrates a baby's social and cognitive development.

American crows, by contrast, are born in full possession of an adult's vocal repertoire, which contains dozens of different calls. They don't yet know, though, what those calls mean or how to use them, and they acquire this knowledge by observing their families. As juveniles, they'll recite the calls, one by one, as if practicing.

Many bird mothers actually start talking to their babies while they're still inside their eggs. This

THE DEVOTION of mother black bears to their cubs (opposite) is an iconic example of parenting. Cubs will stay with their mom for up to two years before striking off on their own.

BEFORE BABY TURKEYS (below) have even hatched, their mothers are already talking to them. For many bird species, this communication helps prepare their offspring for life outside the egg.

so-called perinatal communication helps synchronize hatching, but it may also have a rich social component. Fairy wrens start learning their calls before they've hatched; zebra finches whose parents sing to them in embryo are better able to handle the challenges of adulthood. The same has been reported for turkeys, and some biologists surmise that hatching eggs in incubators causes long-term behavior disorders.

Parental communication is best described among mammals and birds, but they're not the only creatures capable of this. Crocodile mothers are known for devotion to their offspring, and both juveniles and moms vocalize to one another. (Nile crocodile mothers are particularly attentive to calls made by the smallest of their brood.) A few of the meanings are known, including the calls young crocodiles use to solicit help, but most have not yet been translated.

Perhaps that's because people tend to underestimate reptile intelligence. Their nervous systems resemble those of birds, though, and recent research has described a previously unappreciated cognitive complexity. They might possess an equally unappreciated communicative richness. In 2013, when scientists described South American river turtle mothers calling to their young, it was the first record of such behavior in turtles.

As babies grow up, parents teach them how to survive. They might set an example, such as a mother fox who brings her kits a mouse so they can learn to hunt. Or, the instruction can be more active. Mother bears are in constant communication with their cubs, showing them what to eat and what to avoid, passing on lessons they can someday impart to their own children. ◉

FISH DON'T get much credit for being attentive parents, yet many species do care for their young. Among them are clownfish (above); they even care for eggs laid by other clownfish, and tend to the babies as if they were their own.

AS BABIES learn to talk, they go through a period of babbling during which an innate vocal capacity acquires the rhythms of speech. Baby pygmy marmosets (left) babble, too.

GREAT MINDS

» William Long (1867–1952)

In books like *How Animals Talk* and *School of the Woods,* naturalist William Long described animals teaching their young. Long regarded these lessons as yet more evidence of animals as thinking, feeling beings shaped by experience, not just instinct—a view that upset some scientific-minded people, among them President Theodore Roosevelt, who accused Long of sentimentalism and even fraud.

The resulting "nature fakers" controversy was front-page news. Roosevelt's mentality ultimately prevailed—not just in the short-term but for much of the 20th century. With the benefit of hindsight and more research, though, Long now appears as ahead of his time.

CULTURE AND LEARNING

Once considered a defining human characteristic, culture—information that's not inherited at the genetic level but learned from other individuals—has since been described in many animals. Accumulated knowledge, rather than biology alone, is what makes many animals unique.

One of the first scientific accounts of animal cultural transmission came from England, where in 1921 people observed birds in the town of Swaythling prying the tops off milk bottles. Within a few decades the behavior spread across the country. In some places, still more innovations arose, with birds devising new milk-pilfering techniques, or preferring certain grades of milk. The birds had learned from one another, and then built upon that knowledge.

In an ursine echo of those clever avians, a female black bear in upstate New York learned in the mid-2000s to open a state-of-the-art bearproof canister. It was the first such report anywhere in the United States. Before long, other bears in the area adopted her technique, which involved pressing a button, turning the canister's lid, and then pressing a second button.

Did that bear teach the others? Or did they learn by watching her? Nobody knows, but bears—whose intelligence has been likened to that of great apes—are a potentially fertile repository of knowledge. Scientists have observed brown bears in the Russian Far East eating clay, evidently to prevent digestion problems caused by salmon-rich diets. (Geophagy, as it's called, is also practiced by chimpanzees. Along with community-specific tools, nest-building techniques, and even ceremonies, this digestion hack might be a part of their own cultures.)

Bears have certainly influenced human cultures. Eurasian languages contain no fewer than 1,200 plant names that refer to bears—some

VOCALIZATIONS of fruit bats (above) are not a strictly innate affair. They're shaped by social learning, with bat communities possessing unique characteristics that are comparable to human accents.

PERHAPS THE most striking example of the ubiquity of culture among animals is seen in fruit flies (opposite). They can pick up unfamiliar gestures from closely related species, and females even choose mates according to what they've seen other fruit flies doing.

inspired by the plants' appearance, but others by watching bears eat and even self-medicate with them.

Salmon also figure into the culture of an orca clan living off the Pacific Northwest coast, where they possess traditional knowledge of how to find and catch the fish. Examples of cetacean culture abound, from bottlenose dolphins in Australia gathering food with sponges—practices that appear to be taught by mothers to their daughters—to the vocalizations of sperm whales, whose clans possess distinctive dialects.

The same goes for fruit bats whose dialects, learned from other members of their communities, have been analogized to accents. Monk parakeets, a South American native introduced in the 1970s to a handful of cities in the eastern United States, also have calls specific to their communities. These can be seen as sonic testimony of the extent to which they've become locals.

Culture has even been observed in fruit flies, who exchange information via scents and wing movements. Individuals belonging to different fruit fly species initially struggle to communicate—but after a few days of communal living, they readily understand one another's gestures. Females also choose mates according to preferences they've acquired by watching other females, and they in turn set an example for future generations. Though their brains are no larger than poppy seeds, they still have tradition.

PARROTS ARE quick to learn from one another. As conservationists reintroduce endangered scarlet macaws (opposite) to the wild, a newly developed cultural knowledge of landscapes and foods will help them thrive.

GREAT MINDS

» Peter Marler (1928-2014)

When Peter Marler started his research in the early 1950s, portable tape recorders and sonic spectrographs—the latter developed during World War II to represent sound in graphic form—were still newfangled tools. Marler used them to analyze birdsongs, which he discovered were not necessarily hardwired from birth but learned from other birds.

At a time when animal communication was thought to be entirely instinctive, it was a radical idea. If early century naturalists who described animal learning could be dismissed as anecdotalists, Marler made sure to ground his findings in data.

His studies of white-crowned sparrows living around San Francisco Bay showed that songs could vary markedly between populations separated by just a few miles. Not only did the birds learn to sing; their songs were a cultural legacy.

NAMES

Not long after hatching, green-rumped parrotlets learn contact calls from their parents—the sounds they'll use for the rest of their lives to stay in acoustic touch with family and friends.

Each bird learns a call that's unique to him or her. Other parrotlets can mimic the calls and use them to refer to particular individuals. In other words, they have names.

Many other parrot species are believed to share this trait, as do bottlenose dolphins. And animals have many other ways of communicating their identities.

Often their voices are individually distinct. This doesn't qualify as a name, because other animals don't repeat them, but their unique voices let others know who is talking. Consider it a signature. This is especially important for animals who live in groups and also families: shortly after birth, mother Atlantic walruses and their calves recognize one another's voices, and can pick them out from amid the colonial din.

Dwarf mongoose voices are similarly distinctive, although this varies depending on the calls they use. The "isolation calls" made by lost group members to reunite with their clan are individually identifiable. When they're broadcasting warnings about predatory snakes, though, each individual's call sounds pretty much alike. This reflects the principles shaping their communications: Much can

CASE STUDY

»Wolf Howls

A wolf pack's chorus symbolizes wilderness. It also conveys a lot of information.

The choruses contain at least six different types of vocalizations, including barks, whimpers, and howls. Each can be delivered with variation and nuance: Howls prompted by a pack member's absence, for example, sound different from howls spontaneously performed at sunrise.

Scientists haven't yet translated the meanings, but they're clearly essential for coordinating movement between many individuals across vast distances. Howls are also individually distinctive—and so too are the collective sounds of a pack's howls. Just as individual wolves have sonic signatures, so do their tribes.

THAT FEW species use human-style names doesn't mean they don't know one another's identity. For many animals—such as fallow deer (above)—scents are signatures of individual distinction.

be gained by knowing whether a friend or a stranger wants to meet up, but if the message is one of urgent danger, it's not so important to know precisely who sent it.

Many species also possess individually distinctive scents. These can let a deer, for example, know not only that another deer is nearby but also its identity. Even hermit crabs use odor to recognize crabs they've previously met, and combine recognition with memories of past experiences. Complex relationships and social interactions are built on such foundations.

It's a fun thought experiment: What sort of relationships might people have if we could recognize and remember one another but couldn't call one another by name?

Many parrot species, as well as dolphins, use unique calls to distinguish themselves. Animals have many ways of communicating their identities.

SOCIETY

Communication and cognition are intertwined: Powers of thought shape the ability to communicate, which in turn influence the social arrangements in which animals live, which further fuel cognition and communication—and so on across evolutionary time, in multiple feedback loops that have helped shape animal societies.

An especially fertile social arrangement is that of fission-fusion societies, in which groups merge and separate across space and time. A young American crow, for example, spends his first spring with parents and older siblings, whom he leaves in autumn to join far-ranging gangs of fellow juveniles. Each evening they'll join roosts containing thousands of other crows. Come spring he'll return to his family and help raise the next brood.

With that lifestyle comes many benefits. Crows don't always have to find food themselves but can follow their friends. In winter, when they're at risk of attack by hungry raptors, there's safety in numbers; when it's time to raise a family, they have help. The lifestyle brings risks too, particularly the likelihood of conflict. And though crows are not consciously aware of the evolutionary dynamics that select those who thrive at navigating complex social currents, they are products of them.

No wonder that crows, whose cognitive prowess has earned them the moniker of "feathered ape," are so smart. They have a lot to deal with, and their highly developed communication skills, including dozens of calls used in different contexts, are an essential part of that.

THE SIZE and complexity of ant communities have earned them comparisons to human cities. Leafcutter ant colonies (opposite) may contain 10 million individuals, reach dozens of feet into the earth, and sprawl across an acre. Different as we are, we still have some things in common.

GREAT MINDS

» Ben Kilham (b. 1952)

Black bears are often described as solitary creatures, with social contact limited to mother-and-cub interactions and fleeting adult encounters. Ben Kilham, a student of animal behavior and orphaned-cub-rehabber, has spent much of the past two decades studying black bears who live near his New Hampshire home. He's found them to be quite social indeed.

The bears have a hierarchy dominated by an elder female, much like bonobos. Using a communication system that includes context-specific vocalizations, scent deposits, and gestures, they make arrangements and exchange favors that are remembered for years. Far from being solitary, they have society.

Sophisticated cognition and rich communication are found in many fission-fusion species. Among mammals, these include chimpanzees and orangutans and *Homo sapiens*. For us, gossip may—from an evolutionary biologist's perspective—help maintain social cohesion between oft separated group members.

Also living a fission-fusion life are elephants, wild horses, and bottlenose dolphins. The latter have social organizations and communication systems similar to those of many fission-fusion parrot species. That parrots live in treetops and dolphins live underwater is secondary to the social forces that shaped them.

Of course, the fission-fusion lifestyle isn't necessary to be smart and communicative, or to have complex societies. Some ant species live in vast, multiplex colonies that are perhaps nature's closest analogue to modern human cities. They manage this with a few chemical signals and a system of chemically based individual recognition derived from compounds that originally evolved to prevent ants from drying out, and then co-opted to enable solitary-living ants to recognize their kind in order to mate.

At the other end of the size spectrum, orcas and sperm whales are quite intelligent but live in small, stable social groups. Their communications, like those of elephants, coordinate societies whose members are scattered for miles.

Such communication isn't necessarily logistical. For ring-tailed lemurs, it's just about staying in touch. Scientists who study how mutual grooming strengthens lemur relationships say their vocal exchanges serve a similar function. They call it grooming at a distance.

Powers of thought shape the ability to communicate, which in turn shapes the social lives of animals, which further fuels cognition and communication.

LIKE HUMANS, wild horses (above) live in fission-fusion societies. Across days and seasons, herds will split into smaller groups and then join back together.

SOCIAL LIFE helps shape intelligence. The complex social lives of American crows (below), who are sometimes described as "feathered apes," could help explain why they're so cognitively sophisticated.

CASE STUDY

»Sperm Whales

Understanding sperm whales is a challenge: Their activities occur out of human sight, and they regularly journey thousands of miles. Until recently, researchers recording the clicks sperm whales use to communicate had no way to know exactly which whale made them.

By discovering how the shape of each whale's head lends a distinctive sound to their clicks, biologist Hal Whitehead, along with collaborators Luke Rendell, Shane Gero, and Tyler Schulz, made it possible to link calls to individual identities. This revealed that each whale has a unique vocalization that may function as a name, and in turn helped illuminate the family and clan allegiances that shape their societies.

Much remains to be learned, and it won't be easy—but at least scientists will know who's talking.

COMMUNITY DECISIONS

When it's time for a resting pack of African wild dogs to move along, they don't just follow a leader. They make decisions collectively, with consensus emerging after all pack members have had a chance to express their preference. They appear to do this by sneezing. Once a sufficient number of dogs have sneezed, the pack heads off.

Some questions do remain about this sneeze-to-vote system. Perhaps sneezes are not actually votes, but a way of the dogs clearing their noses in preparation for the hunt, and just happen to be made by individuals who want to go. If that's the case, though, the sneezes may follow some other, as yet unidentified way they make their wishes known.

Consensus decision-making systems are found throughout the animal kingdom, from honeybees to swans to elephants. Rarely does one dominant individual decide where everyone else will go. Some may have more influence than others—among those African wild dogs, a high-ranking dog's request requires just three supporting sneezes, while a low-ranking dog's suggestion demands 10 sneezes of support—but they still need community buy-in.

Olive baboons are more likely to follow a union of several

WHEN A PACK of African wild dogs (opposite) is ready to move, individuals vote on where to go. They seem to indicate their preferences by sneezing.

WHOOPER SWANS (below) signal their readiness to depart by moving their heads. When the movements reach a critical frequency, the flock moves. Not everyone will agree with the quorum choice, but being part of a group is often more important than always getting one's own way.

low-ranking individuals than the direction of one dominant baboon. When two members of a European bison herd want to travel in different directions, the herd may briefly separate before reuniting behind one—or they might reach a compromise and follow a trajectory halfway between those suggested.

Leadership can also be fluid. Pigeon flocks don't follow the biggest, strongest pigeon, but instead a pigeon that best knows where to go. They may also choose a new navigator in mid-flight.

Studies of their flight patterns show that larger flocks take longer to settle on a path than do smaller flocks. This is a mathematical signature of participatory decision-making. The more voices there are, the longer it takes—a fair trade, considering that different individuals know different things, and a flock that considers all possibilities can make the most informed decision. This is one benefit of democracy.

Researchers who tracked sperm whale groups found telltale signs of democratic decision-making: slow, messy movements, with individuals pointing different ways before finally settling on a direction. To what extent did they deliberate about possible destinations and reflect upon their choices, as opposed to some simpler mechanism of triangulating between vectors? For now we don't know, and a good scientist would caution against jumping to conclusions. By the same token, it's important to keep an open mind.

Another intriguing report of cetacean behavior involved a community of spotted dolphins who made their ancestral home in the northern Bahamas. Between September 2012 and May 2013, after a prolonged decline in prey, half the dolphins departed and founded a new community some 100 miles away. That unprecedented exodus was no casual thing. It tore apart close families and groups of friends. Scientists don't know if they communicated about it—but it's not difficult to imagine.

AMONG AMERICAN BISON (above), older females tend to determine the direction in which a herd will move, but whether a herd will move in the first place is a community decision.

GREAT MINDS

» Larissa Conradt (b. 1966)

The study of group decision-making goes back millennia. Only recently, though, have scientists examined decision-making in animal collectives. A pioneer in this field is Larissa Conradt, whose early studies of red deer movements turned into a fascination with the principles underlying collective decisions.

One fundamental insight of Conradt's work is that methods for arriving at consensus are found throughout the animal kingdom. And whatever short-term conflicts exist between an individual's desires and those of the group, in the long run it's usually better to maintain social cohesion than insist upon getting one's own way.

INTERSPECIES COMMUNICATION

Open a field guide or zoology textbook and you'll likely find animals arranged by species along taxonomic lines. It's a sensible way of organizing them, but it also tends to shape how we think about their social lives: as confined to members of their own kind, with those taxonomic boundaries crossed only when it's time to eat.

In fact, relationships often transcend species classifications. Songbirds regularly form mixed-species flocks; not only is security found in numbers, but certain species—such as tufted titmice in eastern North America—play the role of sentinels, alerting other birds to threats. Emboldened by their guardians' presence, members of the flock venture into territory they'd avoid on their own. Research shows that chickadees actually understand titmouse warning calls. Woodchucks, though considered a solitary species, likewise tune in to the alarm calls of chipmunks. Mule deer also listen to marmots: They belong to different genera but share a common predator in coyotes.

Sometimes predators work across species lines. Coyotes and badgers have been seen hunting ground squirrels together; perhaps they communicate while on the prowl. Coral trout and coral grouper invite cooperative hunting partners like giant moray eels and big blue octopuses to hunt, and then tell them where to go—something that, until quite recently, was considered beyond their ken.

IMPALAS RELY on oxpeckers (opposite) to eat ticks they otherwise cannot remove from their face and ears. When impalas are ready to be groomed, they will lower their ears or incline their head—signals the oxpeckers understand as invitations to dine.

PISTOL SHRIMP and goby fish (below) depend on each other to survive. The nearly blind shrimp dig burrows in which they live together, and gobies warn their crustacean friends of approaching predators.

ONE REMARKABLE interspecies union is found among goby fish and pistol shrimp. The nearly blind shrimp dig burrows in which they live with gobies; in exchange, the gobies alert them to threats, which they signal with specialized tail flicks. When shrimp leave the hole, they make sure to keep one antenna touching their piscine roommates.

Back on land, salamanders are frequently found cohabitating with rodents in their subterranean burrows. Tiger salamanders reside in prairie dog towns; whether they communicate with those fantastically vocal rodents is unknown, but it's possible. They've certainly had time to figure it out: A 250-million-year-old fossil found in South Africa contained a salamander-like amphibian curled up beside an ancestor of modern mammals.

Interspecies relationships aren't just about survival. The internet abounds with stories of interspecies friendships between domestic and wild animals. Crows have been

IN THE southern Brazilian town of Laguna, bottlenose dolphins collaborate with local artisanal fishers (above). The dolphins herd schools of mullet toward shore, and then signal the fishers to cast nets by slapping their tails on the water; in exchange, they eat some of the fish themselves.

observed playing with cats and dogs, whose play signals—such as a dog's exaggerated belly-down bow—they clearly understand. Fun is universally understood.

So, perhaps, are certain emotions. Mule deer mothers respond to the cries of kittens, baby seals, and even human infants. Studies describe how humans readily identify the sounds of animals in distress—and not just fellow mammals, but birds and amphibians and reptiles. This ability isn't necessarily rooted in some heightened *Homo sapiens* sensitivity, but in fundamental acoustic properties of emotionally charged vocalizations.

We do have some special interspecies relationships, though, both with our companion animals and also with wild creatures. In one southern Brazilian town, artisanal fishers cooperate with certain bottlenose dolphins. The dolphins herd shoals of fish toward shore, and then alert fishers by slapping their tails on the water. After nets have been cast, the dolphins scoop up disoriented fish. The dolphins who do this even have a special whistle; it's not functionally related to fishing but distinguishes them as a separate social group from their noncooperating peers. They too are artisans.

In southern Africa, people have long followed honeyguide birds to beehives. The people have ancient calls used to recruit the birds. Some scientists think this relationship goes back nearly two million years; before humans even existed, the birds helped our ancestors.

It's even been theorized that this ability to find honey, one of nature's most energy-dense foods, played a crucial role in early human development by providing the boost necessary for our exceptionally large, energy-intensive brains to evolve. There's no way of knowing whether this is true, but it's beautiful to think that humans became *human* by learning to understand other animals.

PART THREE

IN TRANSLATION

To appreciate the biological world's communicative richness—to know that a mole cricket tunes his voice to the shape of his burrow, or that a mother deer recognizes the distress in an orphaned kitten's cry, or that a squirrel's waving tail is a signal—is to have a richer sense of life. The landscape outside one's door becomes a place suffused with previously hidden meanings.

Not only are we enriched, but we can also apply our newfound understanding to the relationships we have with other animals: those with whom we share our homes and neighborhoods, and also those impacted by our often unwitting, extra-large human footprint.

Thusly equipped, researchers have found new ways of understanding and making ourselves understood by other creatures. Perhaps these will lead to new forms of coexistence. Even the search for extraterrestrial life could benefit from studies of animal communication. After all, alien communications could well resemble those of dolphins or octopuses rather than our own.

We can better understand how the living world works too, and be cognizant of how our activities affect the ability of animals to communicate. Some, like urban birds who change their songs to be heard above the human din, can adapt—but not all are so fortunate. And at a philosophical level, knowing how animals communicate with one another might also nourish a deeper sense of kinship.

The basis of that kinship is sure to expand. For all that we now know, there's even more to be learned.

THE CHIMPANZEE known as Nim Chimpsky (opposite) was taken from his mother as an infant in the 1970s so psychologists could teach him sign language. The scientific significance of that research—and, more important, its ethics—was highly suspect.

UNDERSTANDING OUR COMPANIONS

Among the traits said to "make us human," our relationships to other animals have received little attention. Yet these relationships have shaped our development as a species, and not just in the sense that domesticating cattle provided a consistent source of protein or riding horses allowed pre-industrial people to travel faster than they could on two feet. Rather, our interdependence with certain other animals has cultivated a deep attunement to them.

Nowadays, those relationships have changed, and few people have the everyday animal experiences our ancestors did. But some animals remain an intimate part of our lives. The journey of understanding them continues.

When people first domesticated dogs, the essential cognitive traits—especially a highly social, communicative nature—of mankind's now best friend were already present. At our side, they've become even more pronounced.

Many studies describe how dogs recognize our emotions and read our body language. So do we read theirs, though subtleties of canine expression, and in particular their movements, may escape even a dog lover's eye.

A wagging tail indeed expresses pleasure, at least when the motion is loose. A stiffer wag, however,

WHEN PEOPLE think about dogs communicating (opposite), they usually think of barks—but body language, from the subtleties of a wagging tail or ear position, may say much more.

GREAT MINDS

» Marc Bekoff (b. 1945)

As an ethologist, Marc Bekoff studies the science of animal behavior. He spent the first part of his career studying the social behaviors of coyotes and other wild dogs. He was fascinated by, among other things, how they played with one another—a seemingly simple activity but founded on a complex communicative back-and-forth and reflecting an innate sense of fairness.

Later Bekoff became a prominent animal advocate, counseling for compassion toward all animals. His original love of dogs remained, though, and he's also a popularizer of scientific research into their behaviors and cognition.

suggests stress. Held high, that wagging tail signifies confidence; drooping low, it suggests anxiety. When dogs feel quite good, their tail wags farther to the right, but confronted with an unfamiliar dog, the tail trends to the left.

Also meaningful are their facial expressions: the set of mouth and eyes and ears. Gesture is so important to dogs that it's almost strange how, when people talk about translating dog communications, they usually focus on decoding their barks.

Compared to dogs, cats are often considered aloof and undemonstrative. That's unfair, though. They just express themselves differently.

Best known are their meows, of course, which are actually a sound they make only infrequently in the wild. Meows are a product of life with us, and they don't have a single meaning. Instead, meows are shaped by a cat's unique interactions with individual humans. In turn, people are better able to identify a recorded meow's context—whether made at a vet's office, say, or before eating—when it was made by their own cat rather than a stranger.

Purrs, by contrast, are universal, though they come in different forms. Low-pitched rumbling signals contentment. Purrs can also be combined with other vocalizations to change their meaning. For example, when a cat is begging for food, they might add a high-pitched sonic layer to their usual low-pitched purr. Combined, these elements are biologically difficult to ignore. They're fashioned to get our attention.

Like dogs and cats, horses are also skilled at reading humans' body language. Less research has been conducted on their communication, though, and indeed on horse welfare in general. Their physical health and social needs are easy to ascertain—but not always their state of mind.

Scientists are now finding ways of assessing horse well-being that can supplement our own often unreliable judgments, such as gauging how horses respond to uncertainty. (If a horse learns that a bucket placed in one location regularly contains good food, for example, and a bucket in another spot contains bad food, how will they react to a bucket halfway between them? Happy horses are more likely to give that uncertain fare a try.) It's an indirect sort of communication, but it might help us better understand them.

HORSES (opposite) have received less scientific attention than cats and dogs, but there's no question that their communicative abilities—with one another and with us—are quite sophisticated.

THE MEOWS of domestic cats (right) are rarely heard in the wild. Instead, they're a signal unique to their relationships with humans, and their meaning evolves through interactions with us.

COMMUNICATING WITH OTHERS

The idea of communicating with other species is a loaded one. People do communicate with animals regularly: Witness our relations with companion dogs and cats, or the intricate coordination of shepherd and sheepdog. Yet the idea of conversing with other animals linguistically, and teaching them to use our languages, is beguiling.

The recent history of this is mixed. Between the late 1960s and early 1980s, scientists attempted to teach sign language to chimpanzees, bonobos, and gorillas. The results and how to interpret them were contested; it's fair to say the apes acquired several hundred words, but they used them with no more sophistication than a young child could.

In hindsight, those experiments feel quite dated. We understand much more now about how great apes communicate with one another, using sounds and gestures particular to their species and environments rather than our own. That they don't readily adapt to human-style language ought not to be a judgment but a statement of fact that doesn't detract from what they are capable of.

Those studies did raise awareness of great ape intelligence, though,

GREAT MINDS

» Denise Herzing (b. 1957)

Since 1985, Denise Herzing's Wild Dolphin Project has followed an Atlantic spotted dolphin community in the Bahamas, meticulously documenting their movements and interactions. It's the longest-running project of its kind.

Of particular interest to Herzing is dolphin communication. She and her collaborators are working with the dolphins to develop a new communication system in which objects—for example, a ball or a piece of seaweed—are associated with whistles that can be mimicked by dolphins or produced by humans using synthesizers. This framework sets an instructive example. Rather than making dolphins talk like humans, we might meet them halfway.

COMMUNICATING with an animal doesn't require a common spoken language. It's an exchange of meaning. Humans can't trumpet to an orphaned African elephant (above), but attentive care is understood.

and even a few signs could be used to great effect. Some of the chimps trained to sign were eventually sold for use in medical experiments; one of the scientists who trained them later visited them at the lab, where one of the chimps reportedly signed to him, "Key. Out."

In a similar vein, researchers funded by the United States Navy and NASA, among others, in the 1950s built an aquarium on the Caribbean island of St. Thomas with the purpose of teaching dolphins to speak English. That program ended in ignominy: The lead scientist also studied LSD's mind-altering powers, and even gave some to the dolphins.

Despite the controversy, it at least drew attention to the impressive intelligence of those cetaceans. As with great apes, we now understand more about how dolphins communicate with one another, and some research continues into how we might speak to the latter (see sidebar opposite) in more natural ways.

Researchers are also developing software to translate the rich vocalizations of prairie dogs. And even without such devices, shared meanings are eminently possible.

AMERICAN CROWS (above) may develop relationships with people who regularly feed them, and may even communicate with their human friends using some of the many vocalizations and gestures that crows use with one another.

KOKO THE GORILLA (opposite) is one of several great apes to whom scientists attempted to teach sign language several decades ago. They didn't use it quite like we do, but that doesn't detract from their ability to communicate—especially when using their own methods, rather than ours.

People who feed crows and ravens develop long-term relationships in which corvids "speak" to them with their species-specific gestures and vocalizations.

Conservationists hoping to attract African elephants to landscapes from which they've been extirpated might recruit them by broadcasting recordings of elephants trumpeting. It's communication as subterfuge. And people who care for orphaned elephants communicate with them. Although a kind voice and gentle hand isn't so imagination-stirring as the thought of translating human language into trumpet blats or infrasonic rumbles, it's far more practical.

That the notion of talking to animals is so fixated on *talking* perhaps reflects our own biases: that communication must take linguistic form to count, and that what is meaningful to humans isn't what we share with other creatures but what we do not. ⦿

That the notion of talking to animals is so fixated on *talking* may reflect our own biases.

NEW TECHNOLOGY

Observing how squirrels respond to recordings of their calls can help scientists understand what they say—but only to a point, as tail waving is an important part of their exchanges. Researchers thus built a robotic squirrel that could wave its own tail. Whether the real squirrels actually thought the robot was one of them, they acted like it.

Such robots are a modern take on the audio playback experiments pioneered shortly after Thomas Edison invented the phonograph (see page 6). Robots have also been used to study the waggle dances of honeybees and the courtship rituals of endangered sage grouse, and to interact with zebrafish.

One can imagine robots becoming even more interactive. Perhaps the software running them could even be designed to mimic the mental processes of animals. Or robots could be used to help protect animals: guiding fish away from polluted areas, or leading threatened birds toward safer migratory paths.

People working in the field of animal-computer interaction are also developing tools that allow other creatures to express themselves, such as musical instruments for grey parrots at an Austrian animal shelter. One instrument works like a disc jockey's turntable; another is played by two parrots in a duet. There are also prototype

AFRICAN GREY PARROTS (above) have a sense of rhythm, and studies show they enjoy some genres of human music more than others. Using custom-designed instruments, African parrots may be able to make new forms of music for themselves.

ROBOTS designed to look and move like animals have been used to study animal behaviors. They might also be used to help animals; robotic fish (opposite) could be programmed to interact with living fish, perhaps guiding them away from danger or pollution.

musical instruments for captive orangutans and elephants.

These projects don't train the animals to produce human-style music for our own listening pleasure. Rather, they're intended to help the animals make sounds that appeal to them. Are those sounds really *music?* It's a hard question to answer, not least because music can be difficult to define, and there's considerable disagreement over whether other animals are truly musical.

In some ways, that debate resembles earlier disagreements over language. It may be settled the same way. Regardless of whether other animals have human-style language, their communications are still full of meaning and intent. And regardless of whether other creatures make music, rhythm and melody are widespread; the cognitive underpinnings of musicmaking—fine-tuned coordination of movement and perception, synchronized social interactions—are highly adaptive.

Theorists can argue about the essence of music. In the meantime, the animals will enjoy themselves.

LOOKING FOR ALIENS

Will humankind ever find intelligent alien life? And if that happens, will we even recognize it?

Perhaps extraterrestrials will make it easy for us. Their signals will cross the Milky Way in a format we're looking for and be as recognizable through our telescopes as Morse code flashing from a spotlight. Or maybe not—in which case, we might take some cues from the intelligent beings with whom we share Earth.

After all, humans are far from alone. Whales had long-distance communication systems long before we even existed. And scientists responsible for analyzing the data gathered by our telescopes and satellites have looked to them, and also to dolphins and squirrel monkeys, for lessons in recognizing alien communications.

The idea isn't that these notably smart, social species have some as yet undeciphered language. Rather, we might find that certain fundamental patterns of human communications are not one-offs but rather reflect principles that transcend species and even galaxies.

One such pattern, known as Zipf's law, describes the frequency with which words are used. If one were to analyze every book ever printed and plot a graph of how often each word appears, it would follow a precise curve that's found in nearly

GREAT MINDS

» Laurance Doyle (b. 1953)

A member of the Search for Extraterrestrial Intelligence Institute's Quantum Astrophysics Group and leader of the NASA team that discovered the first double-star-orbiting planet, Laurance Doyle was among the first astronomers to consider animal signals as a source of insight into aliens.

The "communication intelligence" filters he helped develop draw upon information-structuring patterns common to human and cetacean transmissions. They're especially useful for analyzing signals that don't arrive on the radio frequencies where astronomers have historically focused their search. Outside those channels, a transmission could easily be confused with emanations from interstellar gas or other sources of cosmic radiation. An alien might sound like a collapsing star—but not to an algorithmic ear trained on whale songs.

RESEARCH on the vocalizations of captive bottlenose dolphins (above) found patterns similar to those present in our own communications. They reflect common principles of information transmission and might inform our search for extraterrestrial messages.

every language, from English to Hindi to Japanese.

Information organized this way is exceptionally resilient to interference. Even when some words are lost in transmission, the general meaning is easier to intuit than if all words were equally common. And when researchers with the Search for Extraterrestrial Intelligence (SETI) Institute analyzed the calls of bottlenose dolphins—all of whom, like us, are quite smart and social—they also found that characteristic curve.

Indeed, it seems to be a universal principle of information transmission. Armed with that knowledge, alien seekers can better calibrate the algorithms that pluck signals from the cosmic cacophony. After all, it would be a terrible thing to miss an alien call because it came on an unexpected channel. And who knows: Perhaps we'll find surprising evidence of unexpected intelligence in species we don't now appreciate.

SOUNDSCAPE ECOLOGY

Listen to a chorus of songbirds at dawn—or, better yet, look at a graphical representation of their songs' timing and wavelengths—and something remarkable becomes apparent: Their voices don't overlap on the same frequencies.

Dense as the chorus seems to our ears, each bird's song occupies a unique portion of sonic space. And together with the sounds of other creatures—insects stridulating, frogs calling, coyotes yowling—the whole biophony, as that sonic collective is known, in turn fits with the sound of leaves rustling and water burbling, known as the geophony.

That this fine-tuned order bespeaks not some extraordinary happenstance but the adaptations of creatures maximizing their communicative faculties is known as the niche hypothesis, an insight that is central to the nascent field of soundscape ecology. Unlike traditional bioacoustics, which focuses on how individual animals and species use noise, soundscape ecology studies entire communities.

It treats the biophony as a property that both reflects and shapes life's interactions. Just as ecologists have studied processes like nutrient flows, prey availability, and habitat connectivity, so might they investigate the dynamics of noise. What,

GREAT MINDS

» Bernie Krause (b. 1938)

As an experimental musician in early 1960s San Francisco, Bernie Krause later became a preeminent field recorder, documenting the glorious biological din of habitats from Antarctica to Zimbabwe. Along the way, he developed both the niche hypothesis and the term "biophony," making him a founder of soundscape ecology.

He was also inspired to consider our own sonic niche. The biophony, says Krause, helped shape human music. Early in human history—and presently among some indigenous cultures—human voices and drums fit in nature's chorus rather than drowning it out.

THE BLARING VOICES of great hornbills (above) fit seamlessly into the chorus of their rainforest homes. This is perhaps not a coincidence but reflects a basic principle of ecology.

for example, are the sonic qualities of a healthy fish nursery? How do the acoustic properties of different topographies or vegetation types influence animal activity? When a burned forest begins to regrow, what does that sound like?

Linking soundscapes to measures of biodiversity also gives conservationists a useful new tool. Counting each species by hand or sight can be difficult and time-consuming; it's potentially far easier to put microphones in the field and analyze the recordings. Entire communities can be extrapolated from their sounds.

This provides a new way of assessing human impacts on the rest of nature, both in a given locale and at a more fundamental level. In those dwindling spaces where the human footprint remains light or even nonexistent, the biophony remains full and cleanly arranged. Where human activity is inconsiderate, though, it becomes fragmented and unstructured.

As Earth loses species and populations, life's choruses lose their voices, too. A diminished world is literally a less harmonious place.

DISRUPTION

Appreciating the animal world's informational richness leads inevitably to the question of how human activities affect its transmission.

Noise—from cars and ships and planes and the sheer busyness of 7.5 billion people—has the most obvious impacts. Some animals can adapt to our intrusions into ancestral sonic spaces, as with white-crowned sparrows and Eurasian blackbirds who sing at higher frequencies in cities than in forests. Not all creatures are so fortunate, though.

Songbirds living near highways often have a harder time finding mates than their country brethren. Traffic noise swamps their songs. It also interferes with the echolocation abilities of greater mouse-eared bats, who struggle to catch insects along busy roads. California ground squirrels can't hear their neighbors' alarm calls over the sound of wind turbines.

Noise, light, and chemical pollution all interfere with the ability of animals to communicate.

Noise issues are even worse underwater, where sound travels far and hearing is vital for many creatures. The roar of ship engines is widespread and fast becoming louder in the world's oceans; the most flagrant consequences, such as whales throwing themselves ashore to escape the head-piercing pulses of military sonar, get plenty of attention, but the subtler effects of an ever present din are troublesome too.

Some dolphins have adapted by simplifying their calls. Researchers compare this to the shouted conversations people have in noisy bars—except, of course, dolphins rely on those conversations to survive. Many fish species also struggle amid the uproar: Spawning Atlantic cod and haddock cannot hear the calls of potential mates, and even toadfish, renowned for the loudness of their voices, cannot compete with a vehicular chorus.

These examples are not exhaustive. Rather, they're just a sampling of the noise-affected species scientists study, and just a sampling of the possible impacts yet to be investigated. And disruption isn't limited to hearing.

Agricultural pollutants interfere with the ability of some fish and amphibians to perceive chemical cues. Songbirds exposed to heavy-metal pollution may have smaller vocal repertoires, and sing less frequently, than those living in

NOISE POLLUTION makes it hard for Bryde's whales (above) and many other cetaceans to navigate, find food, and even locate one another.

TO COPE with city noise, house sparrows (opposite) start singing extra early in the morning; those living in especially loud areas have fewer, and smaller, babies.

healthier places. Pollutants can alter the cognitive processes underlying communications.

Even light pollution causes problems, as with frogs who—lacking the safety of darkness—go quiet in their search for mates. And nobody yet knows how human activities distort vital but little-studied forms of communication, such as the detection of ultraviolet light or the ability to perceive and generate electrical signals. Cuttlefish exposed to artificial noise change their colors more frequently, speeding up the skin pattern flashes they use to communicate. It's not clear why

WHITE-RUMPED TANAGERS (above) play the role of sentinels in the mixed-species songbird flocks that travel with them to feed. If that species should decline, the birds who rely on them for predator warnings could also be jeopardized.

POLLUTION isn't just unhealthy; it can also interfere with chemical communication. Even low levels of pollutants can impair the ability of spadefoot frogs (opposite) to recognize their predators' chemical signals.

this happens, but the fact it happens at all is an unsettling sign.

These impacts can have ripple effects. Whales, for example, are considered ecosystem engineers; if they're harmed by noise, other creatures and ecological processes will be affected, too. Meanwhile, the loss of species—not just extinction, but dwindling populations—is itself a form of communicative disruption.

Scientists worry about the loss of birds who serve as sentinels in mixed-species flocks (see page 80); without them, other birds may not fare so well. In Hawaii, where Ōma'o thrush vocalizations combine syllables invented by individual birds with syllables learned from their neighbors, landscape fragmentation leads to a diminishment of their songs. They have fewer birds to learn from.

The loss of animal cultures is especially concerning. Bighorn sheep and many other migratory mammals don't simply rely on an innate sense of direction to guide their voyages, but draw upon a body of knowledge that's passed between animals and across generations. As their populations contract, that knowledge is lost. It may take a very long time to recover.

CHANGING ATTITUDES

Language was once regarded as separating humans from other animals. The study of animal communication blurs, if not downright obliterates, that arbitrary line. So much meaning can be exchanged without using language—a principle that encompasses not just other animals but also the many nonlinguistic forms of communication on which we rely.

To consider communication in its entirety rather than focusing on one narrow linguistic subset illuminates how much *Homo sapiens* and other animals have in common. With that might come a deeper sense of kinship and new kinds of relationships. At a moment of mass extinction and declining populations and dwindling biodiversity, new relationships are certainly needed.

Understanding animal culture yields a deeper sense of how every group and population can be special in its own way. Orcas as a species are not endangered—but some populations, such as the southern resident killer whales of the northeastern Pacific Ocean, are. Their communities are not interchangeable units of population, but the possessors of unique customs and individuals. Protecting them isn't just a matter of preserving genetic diversity. It's about cultural diversity, too.

Learning how animals communicate can also help us interact with them. This has obvious applications in domestic settings—reading your dog's body language can avert problems at the dog park—and with wild animals, too. Black bears are often euthanized after "threatening" people, yet the behavior

RACCOONS (opposite) are among the creatures who thrive in "our" cities—though perhaps cities could be understood as belonging to many species. Some philosophers argue that the ability of animals to communicate should change the way we relate to them.

KINSHIP with cute, cuddly raccoons is a much easier sell than with brown rats (left). Could better appreciating the intimate back-and-forth between baby rats and their devoted mothers make us more tolerant of these oft reviled rodents?

that makes people feel threatened, the so-called bluff charge, is simply part of a series of gestures that bears use to request more space. Often, they're the ones who feel threatened.

Some philosophers see animal communication, and in particular examples of group decision-making (see page 76), as the basis of a conceptual rethink of relations. If animals have their own politics, might it only be fair to find ways of representing them in our own political deliberations? Building a new industrial park or expanding a roadway doesn't just affect species and populations, but also individuals and communities.

It's a complicated issue, to be sure, but it gets at a fundamental question: How can people give voice not just to the voiceless, but to those whose voices are very different from our own? Perhaps these tensions could be resolved more communicatively.

When bison herds threaten to damage crops or wander into roads, for example, people might concentrate on changing the behaviors of the herds' leaders. Cultures of coexistence might also be cultivated: Human-elephant conflicts might be defused by people developing relationships with their local animals, eventually seeding shared norms of acceptable behavior—don't eat crops!—that become part of elephant community custom.

While a few decades ago such ideas would have been framed in terms of conditioning or instilling fear in animals, they can now be seen as a communication challenge. And if at this time we're only able to understand and make ourselves understood by a few species, much more is possible. What might a science of human-animal greetings reveal?

SOME CONFLICTS with wildlife might be averted by finding ways to communicate with them. Could people ever learn to communicate with wolves (opposite) as we do with our dogs?

GREAT MINDS

» Vicki Hearne (1946-2001)

Philosopher, poet, and dog and horse trainer Vicki Hearne rejected the idea that animals should be understood as guided by mechanical instinct and conditioning. This, she said, was superstitious rather than scientific.

Hearne helped defend much maligned pit bulls, challenging the myth of their supposedly innate viciousness. Pit bulls attack people less frequently than do many other dog breeds, argued Hearne; they're just easy to demonize. Hearne also rejected the long-standing idea that horse handling required a dominance founded on coercion. Instead, she urged people to earn the trust and confidence of those powerful, sensitive creatures. The resulting relationships would not be a lesser version of those found between people, but something deep and unique.

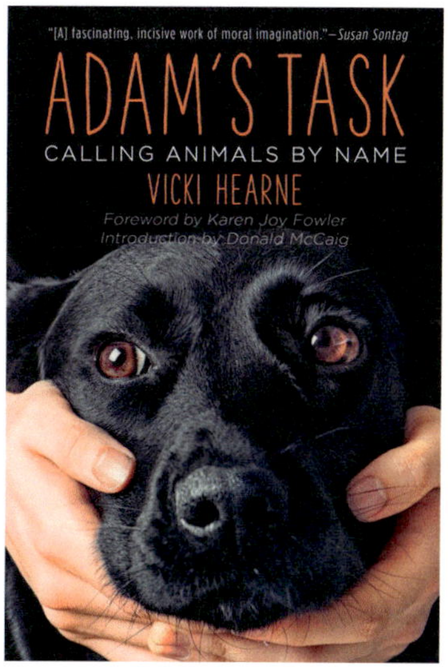

THE UNKNOWN

In the mid-1980s, houseboat dwellers in Sausalito, California, heard a strange humming noise. It was so loud they couldn't sleep at night. Nobody knew what was causing it. The culprit, it turned out, was an aggregation of singing toadfish.

Animal communication is a window into a world of unknowns. Why do birds sing when the sun rises? There are many theories, and the special acoustic properties of predawn air (see page 30) appear to be involved, but there's no one universally accepted explanation. Then there's the question of precisely what they're singing about, which is equally uncertain. Wake up early in springtime and open your window: There sings mystery.

Blue whales are now singing at a noticeably lower pitch than they did just a few decades ago. It might have something to do with noise pollution, or perhaps climate change or shifts in their populations, but nobody is really sure. For that matter, scientists still don't quite understand humpback whale songs. They're extraordinarily well studied, and researchers can track the cultural dynamics shaping their transmission between individuals and clans and across oceans—but as for what the songs actually *do*, uncertainty remains.

So many basic facts have yet to be determined. Ring-tailed coati vocalizations were recently cataloged for the first time. These relatives of raccoons are highly social and cognitively sophisticated, and we've only started to study them. Yet far more is known about coati vocalizations than those of chimney swifts, incredible birds who, in between roosting in old hollow trees or neighborhood chimneys, can fly for months without pause. Even tracing the calls to the individual swift who made them is a challenge.

CACOPHONOUS gatherings of carrion crows (above) are a regular feature of autumn life in many western European cities; the same goes for American crows in North America. What do they talk about when they meet? We don't yet know.

LEOPARD TOADFISH (left) belong to an exceptionally vocal group of fish; they make noise by twitching their swim bladders and are so loud that aggregations can be heard from shore.

CASE STUDY

»Plant Communication

That plants could be likened to animals seems a contradiction in terms. Yet in recent years scientists have described many animal-like patterns of botanical behavior. Plants can share nutrients with needy neighbors, react differently to related plants than they do to those unrelated, and even learn by association.

In an animal, such behaviors would be considered evidence of cognition. Lacking central nervous systems, plants are supposed to be incapable of that—yet when a leaf is damaged, it releases chemical signals that alert other leaves, in a manner not unlike the cells in an animal's body.

Plants may even communicate with one another, perhaps using sound and even magnetic fields as well as chemicals. Incredible—even preposterous—as it seems, we should keep an open mind to the possibility that they're capable of far more than we usually think.

THE COURTSHIP rituals of snapping turtles (above) can last for hours. The turtles will blow streams of bubbles at one another, which some scientists have interpreted as an exchange of chemical information, but nobody knows for sure.

RING-TAILED COATIS (opposite) are quite social and cognitively sophisticated. Their vocalizations were only recently cataloged and, as with so many creatures, have yet to be translated. The study of animal communication shows how much we know—and how much remains to be learned.

Scientists neither know what's going on when courting freshwater turtles blow streams of bubbles at one another, nor the meanings conveyed in the visual grammar of messages that cephalopods flicker across their color-changing, high-resolution skin. Our last common ancestor, a centimeter-long flatworm, swam Earth's oceans some 600 million years ago. How have cephalopod communications—and the mental processes they reflect—converged with those of vertebrates? What about them is unique to their extraordinary evolutionary history?

Scientists don't yet know that, either, but the fact that we can even contemplate communicating with a squid is a marvelous thing. And at a time when it can feel like every inch of Earth's surface has been mapped, that there's nothing left to explore, the study of animal communication is a source of constant delight. It shows just how much we have left to learn.

The journey of discovery will proceed without much of the philosophical baggage that burdened earlier generations. Those old language-based barriers between humans and other animals—the idea that grammar and syntax and symbolism set us unassailably apart—have come tumbling down. Communication doesn't divide us. Rather, it unites us. The message is one of connection and wonder.

SECRETS OF ANIMAL COMMUNICATION
Brandon Keim

PRODUCED BY NATIONAL GEOGRAPHIC PARTNERS, LLC
1145 17th Street NW
Washington, DC 20036-4688 USA

ISSN 2160-7141

Published by Meredith Corporation
225 Liberty Street • New York, NY 10281

Printed in the USA

YOUNG CHEETAHS (opposite) negotiate rough-and-tumble play with gestures that perhaps resemble those of our own distant ancestors.

ILLUSTRATIONS CREDITS

Cover, alyfromuk2us/Getty Images; 0-1, Glenn Jacobs; 2-3, Nori Yuasa/Getty Images; 4, Frans Lanting; 5, GlobalP/Getty Images; 7, Illona Haus/scruffy dog photography; 8, Pete Oxford/Minden Pictures; 10, Joel Sartore/National Geographic Photo Ark, photographed at Sylvan Heights Waterfowl Park; 11, Claus Meyer/Minden Pictures; 12, marcophotos/Getty Images; 13, Eric Isselee/Shutterstock; 14, Eric Isselee/Shutterstock; 14-15, Michio Hoshino/Minden Pictures; 16, Millard H. Sharp/Getty Images; 17 (UP), Matthias Breiter/Minden Pictures; 17 (LO LE and LO RT), Joel Sartore/National Geographic Photo Ark, photographed at Nebraska Wildlife Rehab; 18, Norbert Probst/Getty Images; 19, Otto Plantema/Buiten-beeld/Minden Pictures; 20, David Liittschwager/National Geographic Image Collection; 20-1, Doug Perrine/Nature Picture Library/Getty Images; 22, Nina Leen/The LIFE Picture Collection/Getty Images; 23, Paul Starosta/Getty Images; 24, Tom Murphy/National Geographic Image Collection; 25, Joel Sartore/National Geographic Photo Ark, photographed at the Oklahoma City Zoo; 26, Nora Tam/South China Morning Post via Getty Images; 26-7, Hiroya Minakuchi/Minden Pictures; 28, Bianca Lavies/National Geographic Image Collection; 29 (UP), George Grall/National Geographic Image Collection; 29 (LO), Joel Sartore/National Geographic Photo Ark, photographed at the Fort Worth Zoo; 30, Auscape/UIG via Getty Images; 31, Glenn Bartley/BIA/Minden Pictures; 32, Bird Images/Getty Images; 32-3, Chien Lee/Minden Pictures; 33, Flip Nicklin/Minden Pictures/National Geographic Image Collection; 34, Alan Murphy/BIA/Minden Pictures; 34-5, Sylvain Cordier/Getty Images; 36, Joao P. Burini/NPL/Minden Pictures; 37, dwphotos/Getty Images; 38, Courtesy Toshitaka Suzuki; 38-9, Moonborne/Shutterstock; 40, John Cancalosi/NPL/Minden Pictures; 41 (UP), Lovely Bird/Shutterstock; 41 (LO), Wolfgang Poelzer/Getty Images; 42, Courtesy Mark Pagel; 42-3, Tony Camacho/Science Source; 44, Konrad Wothe/Minden Pictures/National Geographic Image Collection; 46 (LE and RT), Joel Sartore/National Geographic Photo Ark, photographed at the Wellington Zoo; 47, Heidi and Hans-Juergen Koch/Minden Pictures; 48, Robert Seyfarth; 48-9, Neil Aldridge/NPL/Minden Pictures; 50, Jurgen and Christine Sohns/FLPA/Minden Pictures; 51 (UP), James Hager/robertharding/National Geographic Image Collection; 51 (LO), IrinaK/Shutterstock; 52, Courtesy Barbara Smuts; 52-3, Mauricio Handler/National Geographic Image Collection; 54, Tobias Friedrich/Getty Images; 55, Michael Nichols/National Geographic Image Collection; 56, Dickie Duckett/FLPA/Minden Pictures/National Geographic Image Collection; 57, fallbrook/Getty Images; 58, fotolinchen/Getty Images; 59, Bob Elsdale/Getty Images; 60, Peter Chadwick/Getty Images; 60-1, Yva Momatiuk and John Eastcott/Minden Pictures; 62, William Sherman/Getty Images; 63, dssimages/Getty Images; 64, Roland Seitre/Minden Pictures; 64-5, Reinhard Dirscherl/Getty Images; 65, The Book Worm/Alamy Stock Photo; 66, Craig Dingle/Getty Images; 67, David Maitland/Getty Images; 68, OGPhoto/Getty Images; 69, Ingbert Gruttner, Courtesy of the Rockefeller University/Rockefeller Archive Center; 70, Thomas Barrat/Shutterstock; 70-1, Martin-Kubik/Getty Images; 72, Bill Greene/The Boston Globe via Getty Images; 73, Rolf Nussbaumer/NPL/Minden Pictures; 74, yod67/Shutterstock; 74-5, Inga Spence/Getty Images; 75, Brian Skerry; 76, Konrad Wothe/Minden Pictures/National Geographic Image Collection; 77, Paul Souders/Getty Images; 78-9, Charlie Hamilton James/National Geographic Image Collection; 79, Alex Saberi/National Geographic Image Collection; 80, Franco Banfi/Biosphoto/Minden Pictures; 81, Villiers Steyn/Getty Images; 82-3, Timothy Allen/Getty Images; 84, Susan Kuklin/Getty Images; 86, Tom Gordon; 87, Majchy/Getty Images; 88, betyarlaca/Getty Images; 89, Oleksandr Lytvynenko/Shutterstock; 90, Brian Skerry/National Geographic Image Collection; 90-1, Gerry Ellis/Minden Pictures/National Geographic Image Collection; 92, Bettmann/Getty Images; 93, Cheyenne Montgomery/Getty Images; 94, Roslan Rahman/AFP/Getty Images; 94-5, Pete Oxford/Minden Pictures; 96, Courtesy Laurance Doyle; 97, Spencer Lowell; 98, Julien Hekimian/Getty Images; 98-9, BirdHunter591/Getty Images; 100, Aleksey Karpenko/Shutterstock; 100-1, forrest9/Getty Images; 102, Bill Gorum/500px/Getty Images; 103, Glenn Bartley/All Canada Photos/Getty Images; 104, Pierre Aden/Getty Images; 105, Stephen St. John/National Geographic Image Collection; 106, Klein and Hubert/Minden Pictures; 107, Courtesy Skyhorse Publishing; 108, Joel Sartore/National Geographic Photo Ark, photographed at Gulf Specimen Marine Lab and Aquarium; 108-9, Konrad Wothe/NPL/Minden Pictures; 109, Horst Gerlach/Getty Images; 110, Wayne Lynch/All Canada Photos/Getty Images; 111, John Cancalosi/NPL/Minden Pictures; 113, Frans Lanting/National Geographic Image Collection.

Made in the USA
Las Vegas, NV
10 November 2026

57856871R00069